Springer Proceedings in Mathematics & Statistics

Volume 479

This book series features volumes composed of selected contributions from workshops and conferences in all areas of current research in mathematics and statistics, including data science, operations research and optimization. In addition to an overall evaluation of the interest, scientific quality, and timeliness of each proposal at the hands of the publisher, individual contributions are all refereed to the high quality standards of leading journals in the field. Thus, this series provides the research community with well-edited, authoritative reports on developments in the most exciting areas of mathematical and statistical research today.

J. Andrés Christen • Ruth Fuentes-García •
Gabriel Núñez-Antonio • Sergio Pérez •
Alan Riva-Palacio
Editors

Statistics, Society and Environment

35th FNE, Cuernavaca, Mexico, September 27-29, 2023

Springer

Editors
J. Andrés Christen
Department of Probability and Statistics
CIMAT
Guanajuato, Mexico

Gabriel Núñez-Antonio
Department of Mathematics
Universidad Autónoma Metropolitana
Mexico City, Mexico

Alan Riva-Palacio
Department of Probability and Statistics,
IIMAS
UNAM
Mexico City, Mexico

Ruth Fuentes-García
Department of Mathematics, Facultad de
Ciencias
UNAM
Mexico City, Mexico

Sergio Pérez
Department of Socioeconomics, Statistics
and Informatics
COLPOS
Estado de México, Mexico

ISSN 2194-1009 ISSN 2194-1017 (electronic)
Springer Proceedings in Mathematics & Statistics
ISBN 978-3-031-78400-2 ISBN 978-3-031-78401-9 (eBook)
https://doi.org/10.1007/978-3-031-78401-9

Mathematics Subject Classification: 62, 62-06, 62P10, 62P25

This Springer imprint is published by the registered company Springer Nature Switzerland AG
The registered company address is: Gewerbestrasse 11, 6330 Cham, Switzerland

Preface

The National Statistics Forum (Foro Nacional de Estadística, FNE, in Spanish) is the main scientific meeting of the Mexican Statistical Association, taking place yearly, from 1986 to 2018 and every two years since 2018. During 27–29 September 2023, the FNE was held in the Cuernavaca Unit of the Institute of Mathematics (IMATE) of the National Autonomous University of Mexico (UNAM). This was the first in person FNE after the COVID-19 pandemic. Under such context the meeting and present volume focused on the links between statistics, society, and the environment. We thank the local committee members Dr. Carlos Cabrera Ocañas, IMATE-Cuernavaca, UNAM (President), Dr. Aubin Arroyo Camacho, IMATE-Cuernavaca, UNAM, Dr. Ana Lydia Rivera Almaraz, INSP, and M. en C. Gisela Martínez Silva, INSP; the national committee members Dr. Martha María (Mara) Téllez Rojo, INSP (President), Dr. Belem Trejo Valdivia, INSP, and Dr. Carlos Cabrera, IMATE-Cuernavaca, UNAM; and the scientific committee members Dr. Ruth Fuentes, Fac Ciencias UNAM (President), Dr. Manuel Mendoza Ramírez, ITAM, Dr. Eliane Rodrígues, IMATE-CDMX, Dr. Martha Ma. (Mara) Téllez Rojo, INSP, and M. en C. Teresa Verenice Muñoz Rocha, INSP, for all their hard work in organizing the 2023 FNE.

Guanajuato, Mexico — J. Andrés Christen
Mexico City, Mexico — Ruth Fuentes-García
Mexico City, Mexico — Gabriel Núñez-Antonio
Estado de México, Mexico — Sergio Pérez
Mexico City, Mexico — Alan Riva-Palacio

Organization

Program Committee Chairs

Christen, Andrés	Department of Probability and Statistics, CIMAT, Guanajuato, Mexico
Fuentes-García, Ruth	Department of Mathematics, Facultad de Ciencias, UNAM, Mexico City, Mexico

Contents

Toward Inference for Finite Populations

Carlos E. Rodríguez

Abstract This chapter explores three approaches to make inference for finite populations: frequentist, Bayesian, and a novel predictive strategy enforced by martingales. The discussion commences by describing the fundamental concepts underlying sampling techniques. It subsequently differentiates between design-based and model-based inference, emphasizing the introduction of randomness through sampling designs in the former and through probabilistic models describing the random phenomenon of interest in the latter. The predictive paradigm via martingales is explored in the context of missing data, where the observed part of the population is used to construct a predictive model, which is then employed to impute the missing part of the finite population. The chapter concludes by discussing the advantages and disadvantages of each strategy.

Keywords Sampling techniques · Bayesian inference · Predictive modeling · Martingale · Imputation

1 Introduction

Inference concerning finite populations plays a crucial role across diverse fields, encompassing disciplines like medicine, biology, social sciences, and behavioral sciences. Governments and private entities heavily rely on comprehensive data about finite populations, obtained from censuses and surveys, covering aspects ranging from geography and health to economic statistics. This information serves as a cornerstone for policymaking, strategic planning, and decision-making in administration.

C. E. Rodríguez (✉)
Department of Probability and Statistics, Institute for Research in Applied Mathematics and Systems (IIMAS), Universidad Nacional Autónoma de México (UNAM), Mexico City, Mexico
e-mail: carloserwin@sigma.iimas.unam.mx
http://sigma.iimas.unam.mx/carloserwin/

J. A. Christen et al. (eds.), *Statistics, Society and Environment*, Springer Proceedings in Mathematics & Statistics 479, https://doi.org/10.1007/978-3-031-78401-9_1

This chapter aims to elucidate fundamental concepts related to classical sampling techniques. Following this, we delve into the Bayesian perspective on analyzing finite populations, exploring a novel concept centered around predictive inference via martingales. Each approach is meticulously presented, highlighting its distinctive characteristics, along with a thorough discussion on the merits and drawbacks of each methodology. Throughout this exposition, we focus on the estimation of the population mean, $\bar{y}_N$, as a pivotal theme, ensuring clarity and conciseness in presenting each methodology.

2 Sampling Techniques

Sampling techniques (or probabilistic sampling) are some of the most widely used tools in applied statistics. Neyman [14] not only laid the methodological foundations for these techniques but also introduced the concepts of optimal sample allocation and the importance of randomization, which have now become standards in the field. For a detailed account of Jerzy Neyman's general contributions to statistics, including his work on estimation theory, experimental design, and sampling techniques, see [11] (a biographical obituary). This early groundwork was crucial for the widespread adoption of these methods by international agencies for collecting and analyzing data from various countries. Following in these footsteps, William G. Cochran's classic text [1] further elaborated on the theory of probabilistic sampling applied to finite populations, emphasizing practical applications and the development of systematic sampling methods. For a contemporary perspective on these foundational ideas, one can refer to [17], for example.

Let's consider that a certain population of interest has a total of $N < \infty$ elements, denoted by indices $U = \{1, \ldots, N\}$. The objective is to understand a characteristic, Y, of the population U, measured as $y_{1:N} = \{y_1, \ldots, y_N\}$. In this context, $y_{1:N}$ is assumed to be unknown but fixed. Generally, the interest in the characteristic Y is narrowed down to obtaining information measures such as the average $\bar{y}_N = \sum_{i=1}^{N} y_i/N$, total $t_N = N\bar{y}_N$, proportion, or ratio (when there are multiple characteristics of interest). In more general cases, the interest could lie in understanding $\eta_N = \eta(y_{1:N})$, for some function η of $y_{1:N}$. As mentioned earlier, for the sake of simplicity in the explanation, for the rest of the document, we will focus on estimating $\bar{y}_N$.

To determine the population average, $\bar{y}_N$, measuring each element of the population U is necessary. However, this can be costly and time-consuming, especially for large populations or when the measurement process involves complex and/or costly procedures. The solution, provided by the sampling technique, involves: (1) assigning a selection probability $\pi_i = p(z_i = 1) > 0$ for $i = 1, \ldots, N$, known for each element of the population; (2) using these selection probabilities to draw a random sample of n indices from U, where $N \gg n$; (3) measuring the characteristic of interest in the selected population elements, resulting in $y_{1:n}$; (4) using the probabilities π_i to extrapolate the n sample measurements to the population, thus obtaining the estimator $\widehat{\bar{y}}_N$ for $\bar{y}_N$. The random variables z_i indicate whether the ith element of the population is selected in the sample ($z_i = 1$) or not ($z_i = 0$).

It is important to stress that under the sampling technique view, it is assumed, firstly, that the values of the population, $y_{1:N}$, are unknown constants, and secondly, that the observations from the sample, $y_{1:n}$, are measured without any error. Thus, randomness is introduced solely through the random variables $z_{1:N}$, i.e., the strategy by which the information is selected (sampling design). Therefore, probabilistic sampling is known as design-based inference. In contrast, from the Bayesian perspective, all variables are treated as random variables, and a model $p(y|\theta)$ is used, which theoretically describes the random behavior of the sample $y_{1:n}$. This alternative is known as model-based inference and will be discussed later in this chapter.

2.1 Basic Selection Strategies

The basic strategies for selecting elements from the population in probabilistic sampling are simple random sampling without replacement (SRSWOR) and simple random sampling with replacement (SRSWR). SRSWOR is akin to drawing balls from an urn without returning them. In each draw, the urn is thoroughly shuffled, introducing randomness into the process. On the other hand, SRSWR involves drawing a ball, noting its color, placing it back in the urn, and reshuffling before the next draw. This means SRSWR allows for potential repeated selections, while SRSWOR does not. For our description, we will focus on SRSWOR.

Stratified sampling follows SRSWOR, where auxiliary information is used to divide the population of interest into groups called strata. Within each stratum, observations are then selected via SRSWOR. If the behavior of the variable of interest within each stratum is more homogeneous compared to the total population, the estimators for each stratum will have lower variance. Consequently, the overall estimation (weighted average of stratum estimations) will be more precise than in SRSWOR.

In contrast, cluster sampling involves dividing the population into groups (clusters), but instead of selecting within every group, entire clusters are randomly selected. This approach opens the door to multi-stage sampling schemes, where one might perform further sampling within these clusters. Cluster sampling is particularly useful when dealing with geographically dispersed populations, as it reduces costs and logistical complexity. Unlike stratified sampling, where each stratum is known to contribute to the final sample to enhance representation and precision, cluster sampling's focus is often on practicality and operational feasibility.

2.2 Estimation

Once the selection strategy is established, the probability of selecting the ith observation of the population in the sample, π_i, is obtained. In the case of SRSWOR, π_i is given by:

$$\pi_i = P(z_i = 1) = \frac{\binom{N-1}{n-1}}{\binom{N}{n}}, \\ = \frac{n}{N},$$

for $i = 1, \ldots, N$. Therefore, the estimator $\widehat{\bar{y}}_N$ of $\bar{y}_N$ is given by:

$$\widehat{\bar{y}}_N = \frac{1}{N}\sum_{i=1}^{n} w_i y_i = \frac{1}{n}\sum_{i=1}^{n} y_i, \\ = \bar{y}_n,$$

where $w_i = 1/\pi_i$.

Expressing the estimator explicitly as a random variable, i.e.,

$$\widehat{\bar{y}}_N = \frac{1}{n}\sum_{i=1}^{N} z_i y_i,$$

it is straightforward to demonstrate that it is unbiased, obtain its variance, and the estimator of the variance. Assuming asymptotic normality, i.e., for sufficiently large n:

$$\widehat{\bar{y}}_N \sim \mathcal{N}\left(\bar{y}_n, \left(1 - \frac{n}{N}\right)\frac{s_n^2}{n}\right), \tag{1}$$

where s_n^2 is the sample variance.

The general spirit of these methods is non-parametric, as there are no assumptions about the distribution of the variable of interest. While the calculation of point estimators is straightforward, the challenge lies in quantifying the uncertainty of these estimators. For example, in SRSWOR, the variance of the sample mean, $\mathrm{Var}(\bar{y}_n)$, depends on S_N^2, the population variance, which is typically unknown. To address this, we estimate the variance of the estimator using $v(\bar{y}_n)$, which relies on s_n^2, the sample variance. This estimation of variance is crucial for constructing confidence intervals, always under the assumption of asymptotic normality for the sampling distribution of the estimator.

3 Bayesian Approximation to the Analysis of Finite Populations

Bayesian inference in finite populations can be traced back to the pioneering work of [9], where he demonstrated that there are no unbiased estimators of uniformly minimum variance, commonly known as UMVUEs, for the total of a finite population. This revelation sparked renewed interest in analyzing finite populations from a Bayesian perspective. Subsequently, [4] established the connection between exchangeability and sampling from a finite population, introducing the concept of a super-population. This connection laid the groundwork for Bayesian inference on $\bar{y}_N$. For a comprehensive overview of the initial Bayesian proposals addressing the challenges of finite population analysis, refer to [5]. A more general perspective on Bayesian inference in finite populations can be found in [8], while [13] offers an insightful discussion on the evolution of Bayesian ideas in a finite population setting.

To make inferences about $\bar{y}_N$ and stay in line with the sampling techniques approach, it is necessary to use the information $y_{1:n}$ obtained via, for example, SRSWOR. However, under the Bayesian approach, the aforementioned requires additional considerations.

3.1 Super-Population and Finite Population

The first consideration is related to the connection between the finite population $y_{1:N}$ and a super-population $p(y|\theta)$. With this aim [4] assumes that $y_{1:N}$ are exchangeable random variables whose marginal distribution is given by

$$p(y_{1:N}) = \prod_{i=1}^{N} \int p(y_i|\theta)p(\theta)\mathrm{d}\theta, \tag{2}$$

then $y_{1:N}$ can be considered as an independent and identically distributed (i.i.d.) random sample from the super-population $p(y|\theta)$ where the prior $p(\theta)$ incorporates uncertainty regarding θ.

3.2 Working with Samples Obtained via Classic Sampling Techniques

The second consideration aims to conciliate the sample $y_{1:n}$ with the Bayesian framework. We start with the assumption that the *observed data* has been collected from a larger set of *complete data* (or *potential data*), implying the presence of missing data. Considering univariate data, let $y_{1:N}$ be the vector of complete data and $z_{1:N} = (z_1, \cdots, z_N)$ be the observation vector of $y_{1:N}$, where $z_i = 1$ indicates that y_i was observed, while $z_j = 0$ indicates that y_j is missing. Let's introduce the notation in which $y_{1:n} = \{y_i | z_i = 1\}$ corresponds to our observed sample, and $y_{(n+1):N} = \{y_j | z_j = 0\}$ are the unobserved data. It will also be assumed that $z_{1:N}$ is observed (known) in its entirety. These indicator variables are the same random variables used in the sampling techniques section.

From the Bayesian perspective, all variables are random variables, so it is essential to consider the joint distribution of the complete data $y_{1:N}$ and the variables that provide information on how the sample has been selected $z_{1:N}$. The joint distribution that considers a model for data collection must be separated into two parts, one related to the complete data $y_{1:N}$ and the other corresponding to the inclusion $z_{1:N}$. Thus, we can express the *complete likelihood* as the product of the likelihoods of both factors:

$$p(y_{1:N}, z_{1:N} | \theta, \phi) = p(y_{1:N} | \theta) p(z_{1:N} | \phi, y_{1:N}),$$

where θ is the parameter of the distribution of the complete data $y_{1:N}$ and ϕ is a possible parameter of the distribution of $z_{1:N}$. Since we only observe $(y_{1:n}, z_{1:N})$, the appropriate likelihood for making inferences is the *observed likelihood*:

$$\begin{aligned} p(y_{1:n}, z_{1:N} | \theta, \phi) &= \int p(y_{1:N}, z_{1:N} | \theta, \phi) dy_{n+1:N}, \\ &= \int p(y_{1:N} | \theta) p(z_{1:N} | \phi, y_{1:N}) dy_{n+1:N}. \end{aligned}$$

3.3 *Ignorability*

Inference is typically a two-step process: first, making inferences about the super-population through the posterior of the parameters, $p(\theta, \phi | y_{1:n}, z_{1:N})$, and then making inferences about a finite population using the posterior for our missing data $p(y_{n+1:N} | \theta, \phi, z_{1:N})$, through a process called *imputation*. Sometimes, inference is carried out by imputing using the predictive distribution as it allows for comparing different models with different parameterizations.

When the sampling design is unrelated to the variable being estimated, then it holds that

$$p(z_{1:N} | \phi, y_{1:N}) = p(z_{1:N} | \phi),$$

so that the posterior distribution is given by

$$\begin{aligned} p(\theta|y_{1:n}) &\propto p(\theta)p(y_{1:n}|\theta), \\ &\propto p(\theta)\int p(y_{1:N}|\theta)dy_{n+1:N}, \\ &\propto p(\theta)\int p(y_{1:N}, z_{1:N}|\theta,\phi)dy_{n+1:N}. \end{aligned}$$

In this case, the sampling design is said to be *ignorable*, and thus, the posterior and predictive distributions are entirely determined in the usual manner.

The simplest models of data collection are those that are ignorable and known in the sense that:

$$p(z_{1:N}|\phi, y_{1:N}) = p(z_{1:N}), \tag{3}$$

meaning it does not depend on $y_{1:N}$ and there are no unknown parameters ϕ. Thus, only the likelihood $p(y_{1:n}|\theta)$ and the prior $p(\theta)$ are needed for making inferences. In the case of a Simple Random Sampling without Replacement (SRSWOR) where a sample $\mathcal{S}$ of size n is obtained, it is clear that:

$$z_i = \begin{cases} 1, & \text{if } y_i \in \mathcal{S}, \\ 0, & \text{otherwise,} \end{cases}$$

furthermore, if we define:

$$p(z_{1:N}) = \begin{cases} \binom{N}{n}^{-1}, & \text{if } \sum_{i=1}^{N} z_i = n, \\ 0, & \text{otherwise,} \end{cases}$$

then the design is ignorable, making inference straightforward.

3.4 *Putting All the Pieces Together*

First, assuming that $y_{1:N}$ are exchangeable, expression (2), random variables, they can be considered as a random sample from the super-population $p(y|\theta)$. Second, under an ignorable sampling design, such as SRSWOR, expression (3), we can derive the posterior for θ along the usual path, without introducing additional random variables $z_{1:N}$, through $p(\theta|y_{1:n}) \propto p(\theta)p(\theta|y_{1:n})$.

However, a small detail to note is that all of this implies that the sample $y_{1:n}$ consists of i.i.d. random variables, even though SRSWOR does not strictly satisfy the i.i.d. conditions. This approach is justified by the practical notion that, in real-world scenarios, the distinction between SRSWOR and SRSWR becomes negligible, particularly when the population size is significantly larger than the sample size. SRSWR effectively meets the criteria for i.i.d. random variables, making the assumption for SRSWOR practical and applicable in many contexts.

3.5 Estimation

For the simple case of the mean, a widely used approach (see, for example, [7], p. 205) starts with the identity

$$\bar{y}_N = \left(\frac{n}{N}\right)\bar{y}_n + \left(\frac{N-n}{N}\right)\bar{y}_{N-n}, \tag{4}$$

where $\bar{y}_n$ is the observed mean, and $\bar{y}_{N-n}$ is the mean of the unobserved part of the population. To make inferences about $\bar{y}_N$, the predictive distribution is obtained as

$$p(\bar{y}_{N-n}|y_{1:n}) = \int p(\bar{y}_{N-n}|\theta)p(\theta|y_{1:n})d\theta.$$

If $y_{1:n}$ is a random sample from $p(y|\mu,\sigma^2) = \mathrm{N}(y|\mu,\sigma^2)$, with unknown mean and variance, and considering a non-informative prior $p(\mu,\sigma) \propto \sigma^{-2}$. It is straightforward to calculate the predictive distribution for $p(\bar{y}_{N-n}|y_{1:n})$, in closed form, and using the expression (4), we obtain

$$p(\bar{y}_N|y_{1:n}) = t_{n-1}\left(\bar{y}_n, \left(1-\frac{n}{N}\right)\frac{s_n^2}{n}\right), \tag{5}$$

where $t_v(\mu,\sigma^2)$ is a Student's t-distribution with v degrees of freedom, mean μ, and variance $\left(\frac{v}{v-2}\right)\sigma^2$.

The expression (5) is the Bayesian justification for the probabilistic sampling estimator under the assumption of asymptotic normality, i.e., the expression (1).

4 Predictive Inference via Martingales

The methodology described by Fong et al. [6] applies the Bayesian approach to uncertainty quantification, interpreting it in the context of missing data—similar to the approach by Ericson [4], but within a broader context.

Consider an infinite population $y_{1:\infty}$, from which we observe a random sample $y_{1:n}$. The predictive model is then defined as:

$$p(y_{n+1:\infty}|y_{1:n}) = \prod_{i=n}^{\infty} p(y_{i+1}|y_{1:i}),$$

where $p(y_{i+1}|y_{1:i}) = \int p(y_{i+1}|\theta)p(\theta|y_{1:i})\,d\theta$, i.e., the standard Bayesian predictive distribution. Here, a one-step-ahead prediction is made, and the model is updated iteratively: a prediction is generated for the $(n+1)$-th observation, after which the model is updated to predict the $(n+2)$-th, and so forth, effectively imputing the entire population $(y_{1:n}, y_{n+1:\infty})$. The parameter of interest is denoted by $\theta = \theta(y_{1:n}, y_{n+1:\infty})$, which can be precisely determined if the complete population is known.

Realistically, it is impossible to impute an infinite number of observations, $y_{n+1:\infty}$, but we can impute up to $y_{n+1:N}$ for a sufficiently large N. Therefore, with the information $(y_{1:n}, y_{n+1:N})$, the parameter θ is approximated by the statistical functional $\theta_N = \theta(y_{1:n}, y_{n+1:N})$.

From the Bayesian perspective, the posterior mean $\hat{\theta}_N = \mathbb{E}(\hat{\theta}|y_{1:N})$ should intuitively serve as a reliable point estimator for θ. Utilizing martingale theory, as developed by Doob [2], Fong et al. [6] confirms that:

$$\hat{\theta}_N \xrightarrow{a.s.} \theta, \tag{6}$$

where, under some non-restrictive conditions, the essential assumption is that updating $\hat{\theta}_{N-1}$ to $\hat{\theta}_N$ results in a martingale sequence, i.e.,

$$\mathbb{E}(\hat{\theta}_N|y_{1:N-1}) = \hat{\theta}_{N-1}, \quad a.s. \tag{7}$$

ensuring that the imputation is performed only for N large enough to guarantee the convergence of the parameter.

Furthermore, [10] noted that using the posterior mean as a point estimator within Bayesian inference ensures that (7) always holds. However, they demonstrated that Doob's results could also be applied starting from a frequentist estimator. In this case, the predictive model would be expressed as:

$$p(y_{n+1:\infty}|y_{1:n}) = \prod_{i=n}^{\infty} p(y_{i+1}|\hat{\theta}_i).$$

The initial point, $\hat{\theta}_n$, could be the maximum likelihood estimator based on the sample $y_{1:n}$. For $i > n$, $y_{i+1} \sim p(y_{i+1}|\hat{\theta}_i)$, with the update $\hat{\theta}_{i+1} = \psi(\hat{\theta}_i, y_{i+1})$ (for some function ψ), ensuring a martingale, i.e.,

$$\mathbb{E}(\hat{\theta}_{i+1}|y_{1:i}) = \hat{\theta}_i.$$

Additionally, to ensure convergence as stated in (6), it is necessary to demonstrate that the variance, $Var(\hat{\theta}_\infty)$, remains finite. This strategy effectively constitutes a prior-free Bayesian approach.

4.1 Uncertainty Quantification

Thus, in both cases, the missing part of the population is imputed for N sufficiently large to ensure convergence. The algorithms are

Using a Bayesian predictive scheme:

$$
\begin{aligned}
Y_{n+1} &\sim p(y|y_{1:n}),\\
\hat{\theta}_{n+1} &= \mathbb{E}(\theta|y_{1:n+1}),\\
Y_{n+2} &\sim p(y|y_{1:n+1}),\\
\hat{\theta}_{n+2} &= \mathbb{E}(\theta|y_{1:n+2}),\\
&\vdots\\
Y_N &\sim p(y|y_{1:N-1}),\\
\hat{\theta}_N &= \mathbb{E}(\theta|y_{1:N-1}),
\end{aligned}
$$

Via an initial estimate $\hat{\theta}_n$ of θ:

$$
\begin{aligned}
Y_{n+1} &\sim p(y|\hat{\theta}_n),\\
\hat{\theta}_{n+1} &= \psi(\hat{\theta}_n, y_{n+1}),\\
Y_{n+2} &\sim p(y|\hat{\theta}_{n+1}),\\
\hat{\theta}_{n+2} &= \psi(\hat{\theta}_{n+1}, y_{n+2}),\\
&\vdots\\
Y_N &\sim p(y|\hat{\theta}_{N-1}),\\
\hat{\theta}_N &= \psi(\hat{\theta}_{N-1}, y_N),
\end{aligned}
$$

Here we would need to ensure that $\mathbb{E}(\hat{\theta}_{i+1}|y_{1:i}) = \widehat{\theta}_i$, and also $Var(\hat{\theta}_\infty) < \infty$.

Either of these algorithms would produce one complete imputed population and an estimate $\hat{\theta}_N$. However, to perform uncertainty quantification we need to impute B complete populations while obtaining $\hat{\theta}_N^1, \hat{\theta}_N^2, \ldots, \hat{\theta}_N^B$. The corresponding quantiles of this random sample can be used to quantify the uncertainty about the estimation of θ.

4.2 The Bootstrap

The algorithms described in the previous section indeed represent Bayesian Bootstraps. To understand this, we first describe the classic Bootstrap method introduced by Efron [3], followed by the fundamental ideas under Finite Population Bootstraps described by Lo [12].

The classic Bootstrap aims to estimate a statistical functional such as

$$\theta = T(F) = \int g(y) dF_Y(y).$$

If $Y_1, Y_2, \ldots, Y_n$ is a random sample from $Y \sim F_Y(y)$, then we use the empirical distribution function

$$\hat{F}_n(y) = \frac{1}{n} \sum_{i=1}^{n} \mathbb{I}_{\{Y_i \leq y\}},$$

to approximate $F_Y(y)$, making it straightforward to obtain the plug-in estimate of θ via

$$\hat{\theta}_n = T(\hat{F}_n) = \frac{1}{n}\sum_{i=1}^{n} g(Y_i).$$

To approximate the sampling distribution of $\hat{\theta}_n$, the Bootstrap proposes a resampling procedure based on the original sample to simulate the estimator's variability. The method involves repeatedly generating random samples from the empirical distribution $\hat{F}_n(y)$, each of these samples is called a "bootstrap replicate." By repeating this process a large number of times (e.g., B times), an empirical distribution of $\hat{\theta}_n^*$ is obtained, which is used to estimate the distribution of the original estimator $\hat{\theta}_n$. The algorithm is described as follows.

$$\begin{array}{ccc}
Y_1^{1*}, Y_2^{1*}, \ldots, Y_n^{1*} \overset{\text{iid}}{\sim} \hat{F}_n(y), & \rightarrow & \hat{\theta}_n^{1*}, \\
Y_1^{2*}, Y_2^{2*}, \ldots, Y_n^{2*} \overset{\text{iid}}{\sim} \hat{F}_n(y), & \rightarrow & \hat{\theta}_n^{2*}, \\
\vdots \quad \vdots \quad \vdots \quad \vdots \quad \vdots & & \vdots, \\
Y_1^{B*}, Y_2^{B*}, \ldots, Y_n^{B*} \overset{\text{iid}}{\sim} \hat{F}_n(y), & \rightarrow & \hat{\theta}_n^{B*}.
\end{array}$$

Hence, the bootstrap estimates $\hat{\theta}_n^{1*}, \hat{\theta}_N^{2*}, \ldots, \hat{\theta}_N^{B*}$ are used to approximate the sampling distribution of $\hat{\theta}_n$.

Departing from [3, 12] describes the central ideas under finite population Bootstraps in both settings. Starting from $y_{1:n}$, an SRSWOR from a finite population $y_{1:N}$, we are interested in the finite population parameter

$$\eta_N = \eta(y_{1:N}).$$

Thus, the missing part of the population is imputed via

$$\begin{aligned}
Y_{n+1} &\sim \hat{F}_n(y), \\
Y_{n+2} &\sim \hat{F}_{n+1}(y) = \frac{1}{n+1}\left(n\hat{F}_n(y) + \mathbb{I}_{\{Y_{n+1} \le y\}}\right), \\
Y_{n+3} &\sim \hat{F}_{n+2}(y) = \frac{1}{n+2}\left((n+1)\hat{F}_{n+1}(y) + \mathbb{I}_{\{Y_{n+2} \le y\}}\right), \\
&\vdots \quad \vdots \quad \vdots \quad \vdots \quad \vdots \quad \vdots \quad \vdots \\
Y_N &\sim \hat{F}_{N-1}(y) = \frac{1}{N-1}\left((N-2)\hat{F}_{N-2}(y) + \mathbb{I}_{\{Y_{N-1} \le y\}}\right),
\end{aligned}$$

once the finite population is imputed, to perform Bayesian inference we calculate

$$\hat{\eta}_N = \eta(y_{1:n}, Y_{n+1:N}),$$

and uncertainty quantification is done by inputting B complete populations, thus obtaining $\hat{\eta}_N^1, \hat{\eta}_N^2, \ldots, \hat{\eta}_N^B$. The corresponding quantiles of this random sample are used to quantify the uncertainty about the estimation of η_N. This is a Finite Population Bayesian Bootstrap (FPBB). The aim under the FPBB is to approximate the predictive distribution $p(\eta_N | y_{1:n})$, while frequentists would seek to obtain the sampling distribution of an estimator.

Following frequentist ideas, first, we need an estimator for η_N based on the sample $y_{1:n}$, say $\hat{\eta}_n$. Once the imputed finite population $\{y_{1:n}, Y_{n+1:N}\}$ is obtained, a SRSWOR $y_{1:n}^*$ is selected and then the estimator $\hat{\eta}_n^{*1}$ is calculated. This process is repeated B times to obtain $\hat{\eta}_n^{*1}, \ldots, \hat{\eta}_n^{*B}$, and these are used to approximate the sampling distribution of $\hat{\eta}_n$.

These are the fundamental ideas underpinning both Bayesian and frequentist Bootstrap methods for finite populations. To conclude this section, it is worth mentioning that Lo described the FPBB in terms of a Pólya sample of size $N - n$ drawn from the data urn $y_{1:n}$. Our description, while framed differently, is essentially equivalent and leads to an important conclusion regarding the update of the empirical distribution function. Specifically, this update process can be characterized as a martingale, as shown by the following relationship:

$$\begin{aligned}
\mathbb{E}(\hat{F}_{i+1}(y) \mid y_{1:i}) &= \frac{1}{i+1}\left(i\hat{F}_i(y) + \mathbb{E}(\mathbb{I}_{\{Y_{i+1} \leq y\}})\right), \\
&= \frac{1}{i+1}\left(i\hat{F}_i(y) + \hat{F}_i(y)\right), \\
&= \hat{F}_i(y).
\end{aligned}$$

This demonstrates that each update to the empirical distribution function maintains its expected value, affirming the martingale property.

5 Finite Population Parametric Bayesian Bootstrap

Returning to the martingale predictive ideas discussed by Fong et al. [6] and Holmes and Walker [10], as described in Sect. 4, this discussion applies within two specific constraints: (1) a finite population of size $N < \infty$, and (2) analysis under the framework of Finite Population Bayesian Bootstraps (FPBB). From this perspective, several critical insights emerge:

Firstly, within this finite framework, the convergence typically described by Doob's theorem is not required. Monitoring convergence is unnecessary as the value of N is predetermined by the finite population size. Secondly, from a frequentist perspective, although defining a martingale to update the parameter remains challenging, it is no longer essential to demonstrate that $V(\hat{\theta}_\infty) < \infty$. Finally, in a finite population context, the focus may shift from the model parameter θ to a specific finite population parameter $\eta_N = \eta(y_{1:N})$.

The algorithms outlined in Sect. 4.1 truly exemplify Finite Population Parametric Bayesian Bootstraps. Utilizing any of these algorithms to impute a complete population allows us to proceed as specified in FPBB to obtain an estimate $\hat{\eta}_N = \eta(y_{1:n}, Y_{n+1:N})$ for η_N. Uncertainty quantification is then achieved by simulating B complete populations, resulting in estimates $\hat{\eta}_N^1, \hat{\eta}_N^2, \ldots, \hat{\eta}_N^B$.

These methodological modifications enhance the accuracy of predictions and assessments of the uncertainty associated with η_N, leveraging the structured approach of FPBB without the constraints imposed by traditional statistical convergence requirements.

5.1 Estimation

Continuing with the goal of making inference for the mean of a finite population, $\bar{y}_N$, predictive approaches via martingales are explored, incorporating both Bayesian and frequentist philosophies. Assuming a model $N(y|\mu, \sigma^2)$, a reference prior $p(\mu, \sigma^2) \propto \sigma^{-2}$ and leveraging Bayesian updating techniques, the predictive value for y_{i+1}, for $i = n, \ldots, N-1$, is specified by

$$y_{i+1} \sim p(y_{i+1} \mid y_{1:i}) = t_{i-1}\left(\bar{y}_i, \left(1 + \frac{1}{i}\right) s_i^2\right),$$

which can be efficiently executed.

For frequentist initial estimates, the classical maximum likelihood estimates (MLE) for the mean and variance, $\hat{\mu}_n = \bar{y}_n$ and $\hat{\sigma}_n^2 = \frac{1}{n}\sum_{i=1}^{n}(y_i - \bar{y}_n)^2$, serve as the starting points. The updates for $i = n, \ldots, N-1$, are given by

$$y_{i+1} \sim p(y_{i+1} \mid \hat{\mu}_i, \hat{\sigma}_i^2) = N(y_{i+1} \mid \hat{\mu}_i, \hat{\sigma}_i^2),$$

where the updates are formulated as follows:

$$\hat{\mu}_{i+1} = \frac{1}{i+1}\left(i\hat{\mu}_i + y_{i+1}\right), \quad \text{and}$$

$$\hat{\sigma}_{i+1}^2 = \frac{1}{k_{i+1}}\left(k_i\hat{\sigma}_i^2 + \frac{i+1}{i+2}(\hat{\mu}_i - y_{i+1})^2\right),$$

with $k_i = \sum_{j=1}^{i} \frac{j}{j+1}$. These parameter updates define martingales, illustrating a straightforward extension of the maximum likelihood estimators for mean and variance.

Upon obtaining the imputed population (following either approach), i.e., $\{y_{1:n}, Y_{n+1:N}\}$, an estimate of the population mean $\bar{y}_N$ is readily computed. This synthesis of Bayesian and frequentist methods provides a robust framework for estimating population parameters under finite conditions.

6 Example

A simple simulated example has been designed to demonstrate the performance of the methods described in this chapter. We generated a fictitious population of votes cast in favor of two candidates (Y_1, Y_2) across $N = 300$ voting ballots using the following process: For each ballot i, from $i = 1$ to 300, we generated votes as follows:

$$\theta_{i,1} \sim \text{Beta}(0.5, 0.5) \Rightarrow y_{i,1} \sim \text{Bin}(20, \theta_{i,1}),$$
$$\theta_{i,2} \sim \text{Beta}(4, 6) \Rightarrow y_{i,2} \sim \text{Bin}(100, \theta_{i,2}).$$

The objective is to estimate the percentage of votes cast in favor of each candidate, given by:

$$p_j = 100\frac{\sum_{i=1}^{N} y_{i,j}}{\sum_{j=1}^{2}\sum_{i=1}^{N} y_{i,j}} = 100\frac{\bar{y}_{N,j}}{\sum_{j=1}^{2}\bar{y}_{N,j}}, \quad \text{for } j = 1, 2. \tag{8}$$

Given an SRSWOR of size n, expression (8) indicates that estimating population percentages can be simplified to modeling means. Thus, we can straightforwardly use expressions (1) and (5) to generate interval estimates via frequentist and Bayesian approaches, respectively, using simulation. To generate these estimates, we created B simulated mean population vote estimates, from which we directly derived B estimates of the population percentages by taking the corresponding quantiles. For the Finite Population Parametric Bayesian Bootstrap (FPPBB), the entire populations were imputed B times. With these complete populations, we trivially calculated B estimates of the population percentages and then simply took the corresponding quantiles.

To empirically test each strategy, $m = 200$ samples, each of size $n = 10$, were selected. For each sample, interval estimates with 95% confidence or probability were calculated for (p_1, p_2). We monitored the percentage of intervals that contained the true values. The results are presented in the following Table 1.

In this case, all methods demonstrate good performance, though the Bayesian alternatives achieve marginally higher true coverage percentages. For the sake of reproducibility, the relevant code is provided in the appendix.

Table 1 True coverage percentage

Method	p_1	p_2
Sampling	94.0	94.0
Bayesian	95.5	95.5
FPPBB_1	95.5	95.5
FPPBB_2	94.0	94.0

7 Discussion

Having closed-form expressions for the point estimator and an approximation for the variance facilitates straightforward, quick, and reliable interval estimation through classic sampling techniques. When assumptions of asymptotic normality hold and outliers are absent, this is generally the recommended approach.

However, the sampling distribution of an estimator does not always conform to a normal distribution; it can be skewed, possess heavy tails, or be contaminated by outliers. These issues are prevalent in cases where the sample size is "small" or when outliers are present. Even if a normal distribution seems a suitable approximation for the sampling distribution of the estimator, practical problems often arise. For instance, if the inference goal is interval estimation, such as for $\bar{y}_N$, the actual coverage of asymptotic normal intervals may fall significantly short of the designed confidence level. This discrepancy poses a serious methodological and practical challenge, underscoring the need for alternative methods that provide confidence intervals with accurate coverage without relying on the assumption of asymptotic normality. Wilcox [19] offers an extensive discussion on this topic.

The Bayesian alternative, which quantifies uncertainty through predictive distributions based on a model, offers considerable flexibility in addressing diverse problems. The efficacy of this approach heavily depends on the model's capacity to accurately represent the underlying random phenomenon. Incorporating prior knowledge about the variable of interest through the prior is possible, presenting distinct conceptual advantages. For instance, the likelihood principle asserts that once a probability model is established, all inference should be derived from the likelihood, a principle to which Bayesian methods adhere. However, this is contrasted in probabilistic sampling, where inference relies on selection probabilities, thus deviating from the likelihood principle.

Obtaining closed-form expressions for the predictive distribution, e.g., $p(\bar{y}_N|y_{1:n})$, when estimating the population mean, is feasible only with trivial models $p(y|\theta)$ and specific prior choices, such as conjugate priors. This limitation is a clear disadvantage because it generally necessitates the use of computationally intensive methods, like Markov Chain Monte Carlo (MCMC) methods [15], to approximate both the posterior and predictive distributions. This complexity escalates when the target of inference extends beyond $\bar{y}_N$ to more general population parameters η_N.

When the goal is to estimate population parameters beyond simple statistics like means, totals, or ratios—such as minimums, maximums, or quantiles—resampling methods such as Bootstrap [3] or cross-validation techniques [16] become crucial. Nevertheless, adapting these methods to finite population analysis often necessitates significant modifications, as explored in the foundational ideas of [12] and recent discussions like [18] focused on Bootstrap adaptations.

In practice, while the martingale-enforced predictive strategy for estimating parameters like $\bar{y}_N$ is computationally demanding, simpler strategies might be preferred. Nevertheless, parallel processing can alleviate this computational burden, similar to any Bootstrap strategy. Furthermore, the Finite Population Bayesian Parametric Bootstrap approach is especially pertinent in scenarios with small sample sizes and the presence of outliers. Additionally, adopting a frequentist starting point within this strategy allows for straightforward handling of various models, such as finite mixture models, without resorting to computationally intensive techniques like MCMC, essential in many Bayesian analyses. Importantly, estimating any general parameter η_N becomes straightforward through $\hat{\eta}_N = \eta(y_{1:N}, Y_{n+1:N})$ within this alternative framework.

Appendix

Code used for the simulations.

```
# SIMULATED FINITE POPULATION
N <- 300
theta1N <- rbeta(N, 0.5, 0.5)
theta2N <- rbeta(N, 4, 6)
y1N <- rbinom(N, size = 20, prob = theta1N)
y2N <- rbinom(N, size = 100, prob = theta2N)
POB <- data.frame(y1 = y1N, y2 = y2N)
# POPULATION PARAMETER
tot <- colSums(POB)
p <- 100*tot/sum(tot)

# FINITE POPULATION PARAMETRIC BAYESIAN BOOTSTRAPS
normal_FPPBB <- function(n, N, yn, a, B){
  mun <- colMeans(yn)
  sign <- apply(yn, 2, sd)
  an <- colSums(yn)
  bn <- colSums(yn^2)
  yNhat <- array(NA, c(B, 2))
  y2Nhat <- array(NA, c(B, 2))
  kn <- sum((1:n)/(2:(n+1)))
  yNstar <- array(NA, c(N, 2))
  y2Nstar <- array(NA, c(N, 2))
  for (j in 1:B) {
    yNstar[1:n, ] <- as.matrix(yn)
    y2Nstar[1:n, ] <- as.matrix(yn)
    mui <- mun
    sigi <- sign
    ai <- an
    bi <- bn
    mu2i <- mun
    s2i <- (bn - n*mun^2)/(n-1)
    ki <- kn
    for (i in n:(N-1)) {
      yNstar[i+1, ] <- rnorm(2, mui, sigi)
      y2Nstar[i+1, ] <- mu2i + rt(2, df = i - 1)*sqrt((1 + 1/i)*s2i)
      #
      kip1 <- ki + (i + 1)/(i + 2)
      sigi <- sqrt((1/kip1)*(ki*sigi^2 + ((i + 1)/(i + 2))*(mui - yNstar[i+1, ])^2))
      mui <- (1/(i + 1))*(i*mui + yNstar[i+1, ])
      ai <- ai + y2Nstar[i+1, ]
      bi <- bi + y2Nstar[i+1, ]^2
      mu2i <- ai/(i + 1)
      s2i <- (bi - (i + 1)*(mu2i^2))/i
```

```
    }
    yNhat[j, ] <- colMeans(yNstar)
    y2Nhat[j, ] <- colMeans(y2Nstar)
  }
  list(inorm = apply(100*yNhat/rowSums(yNhat), 2, quantile, c(a/2, 1 - a/2)),
       ipred = apply(100*y2Nhat/rowSums(y2Nhat), 2, quantile, c(a/2, 1 - a/2)))
}

between <- function(x, myINT){
  a <- myINT[1]
  b <- myINT[2]
  z <- 0
  if(x <= b & x >= a){
    z <- 1
  }
  z
}

# GLOBAL PARAMETERS
a <- 0.05
n <- 10
z <- qnorm(1 - a/2)
tt <- qt(1-a/2, df = n-1)
B <- 1000
m <- 200

# TO MONITOR THE TRUE COVERAGE PERCENTRAGE
n1 <- c(0, 0)
n2 <- n1
n3 <- n1
n4 <- n1
for (h in 1:m) {
  # SELECTING THE SAMPLE
  id <- sample.int(N, size = n, replace = FALSE)
  yn <- POB[id, ]
  ##############
  # ESTIMATION #
  ##############
  mun <- colMeans(yn)
  sign <- apply(yn, 2, sd)
  sn <- sqrt((1 - n/N)/n)*sign
  # FREQUEINTIST ESTIMATION OF THE PROPORTIONS VIA MODELLING OF THE MEANS
  res11 <- matrix(rnorm(2*B, mun, sn), nrow = B, ncol = 2, byrow = TRUE)
  res12 <- apply(100*res11/rowSums(res11), 2, quantile, c(0.025, 0.975))
  # BAYESIAN ESTIMATION OF THE PROPORTIONS VIA MODELLING OF THE MEANS
  res21 <- matrix(rep(mun, B), nrow = B, byrow = TRUE) + t(matrix(rt(2*B, df = n-1), nrow = 2, ncol = B,
        byrow = TRUE)*sn)
  res22 <- apply(100*res21/rowSums(res21), 2, quantile, c(0.025, 0.975))
  #
  n1[1] <- n1[1] + between(p[1], res12[,1])
  n1[2] <- n1[2] + between(p[2], res12[,2])
  #
  n2[1] <- n2[1] + between(p[1], res22[,1])
  n2[2] <- n2[2] + between(p[2], res22[,2])
  # FPPBB
  res <- normal_FPPBB(n, N, yn, a, B)
  n3[1] <- n3[1] + between(p[1], res$inorm[,1])
  n3[2] <- n3[2] + between(p[2], res$inorm[,2])
  #
  n4[1] <- n4[1] + between(p[1], res$ipred[,1])
  n4[2] <- n4[2] + between(p[2], res$ipred[,2])
}
res <- 100*matrix(c(n1, n2, n3, n4), ncol = 2, byrow = TRUE)/m
colnames(res) <- c("y1", "y2")
rownames(res) <- c("Freq", "Baye", "FPPBB_1", "FPPBB_2")

library(xtable)
latex_table <- xtable(res, caption = "True Coverage Percentage", label = "tab")
print(latex_table, include.rownames = TRUE, include.colnames = TRUE,
      floating = TRUE, hline.after = c(-1, 0, nrow(res)), comment = FALSE)
```

References

1. Cochran, W. (1977). *Sampling Techniques*, A Wiley publication in applied statistics, 3rd edn, Wiley India Pvt. Limited.
2. Doob, J. L. (1949). Application of the theory of martingales, *Actes du Colloque International Le Calcul des Probabilite's et ses applications*, Paris CNRS, Lyon, pp. 23–27.

3. Efron, B. (1979). Bootstrap Methods: Another Look at the Jackknife, *The Annals of Statistics* **7**(1): 1–26.
4. Ericson, W. A. (1969). Subjective bayesian models in sampling finite populations, *Journal of the Royal Statistical Society. Series B (Methodological)* **31**(2): 195–233.
5. Ericson, W. A. (1988). 9 bayesian inference in finite populations, *in* P. Krishnaiah and C. Rao (eds), *Sampling*, Vol. 6 of *Handbook of Statistics*, Elsevier, pp. 213–246.
6. Fong, E., Holmes, C. and Walker, S. G. (2022). Martingale posterior distributions, https://arxiv.org/abs/2103.15671. to appear in Journal of the Royal Statistical Society, Serics B.
7. Gelman, A., Carlin, J., Stern, H., Dunson, D., Vehtari, A. and Rubin, D. (2013). *Bayesian Data Analysis*, Chapman & Hall/CRC Texts in Statistical Science, 3rd edn, Taylor & Francis.
8. Ghosh, M. and Meeden, G. (1997). *Bayesian Methods for Finite Population Sampling*, Routledge.
9. Godambe, V. P. (1955). A unified theory of sampling from finite populations, *Journal of the Royal Statistical Society. Series B (Methodological)* **17**(2): 269–278.
10. Holmes, C. C. and Walker, S. G. (2023). Statistical inference with exchangeability and martingales, *Philosophical Transactions of the Royal Society A: Mathematical, Physical and Engineering Sciences* **381**(2247): 1–17.
11. Lehmann, E. L. (1994). Jerzy neyman, 1894–1981, *The Annals of Statistics* **22**(3): 1677–1682.
12. Lo, A. Y. (1988). A Bayesian Bootstrap for a Finite Population, *The Annals of Statistics* **16**(4): 1684–1695.
13. Mendoza, M., Contreras-Cristán, A. and Gutiérrez-Peña, E. (2021). Bayesian analysis of finite populations under simple random sampling, *Entropy* **23**(3).
14. Neyman, J. (1934). On the two different aspects of the representative method: The method of stratified sampling and the method of purposive selection, *Journal of the Royal Statistical Society* **97**(4): 558–625.
15. Robert, C. P. and Casella, G. (2005). *Monte Carlo Statistical Methods (Springer Texts in Statistics)*, Springer-Verlag, Berlin, Heidelberg.
16. Stone, M. (1974). Cross-validatory choice and assessment of statistical predictions, *Journal of the Royal Statistical Society. Series B (Methodological)* **36**(2): 111–147.
17. Tille, Y. (2020). *Sampling and Estimation from Finite Populations*, Wiley Series in Survey Methodology, Wiley.
18. Wang, Z., Peng, L. and Kim, J. K. (2022). Bootstrap Inference for the Finite Population Mean under Complex Sampling Designs, *Journal of the Royal Statistical Society Series B: Statistical Methodology* **84**(4): 1150–1174.
19. Wilcox, R. (2017). *Introduction to Robust Estimation and Hypothesis Testing*, Statistical Modeling and Decision Science, 3rd edn, Elsevier Science.

A Splitting Criterion for CART Models Based on Bayesian Optimisation

Erick G. G. de Paz, Humberto Vaquera Huerta, Francisco Javier Albores Velasco, John R. Bauer Mengelberg, and Juan Manuel Romero Padilla

Abstract Classification and Regression trees (CARTs) are prediction models based on local estimators assigned to non-overlapping subsets of the data space. To compute a CART model, splitting a hyperrectangular set into two subsets is a recurrent operation defined by the axis to split and the splitting proportion. The optimal axis and proportion should minimise the squared difference between the observed data and the local estimators. In optimisation terms, for any given axis, the squared difference is an objective function of the splitting proportion. In this paper, a Bayesian optimisation technique is applied to this objective function by means of a bound estimation based on a non-informative prior. The estimated bound is used as an acquisition function to search the optimal splitting proportion. Under challenging conditions, several Monte Carlo simulations show that this approach is statistically more accurate and robust in high dimensions than regular CART models.

Keywords CART models · Bayesian optimisation · High-dimensional data · Machine learning

1 Introduction

Classification and Regression Trees (CARTs) are a family of machine-learning models for approximation and prediction [1]. The construction of a continuous CART model is sequentially guided by the criterion of minimising a frequentist point estimation of the mean squared error [2]. In this chapter, a Bayesian optimisation technique is applied to estimate and then minimise the error of the model by

E. G. G. de Paz (✉) · H. Vaquera Huerta · J. R. B. Mengelberg · J. M. R. Padilla
Colegio de Posgraduados, Texcoco, Mexico
e-mail: giles.erick@colpos.mx; hvaquera@colpos.mx; jbauer@colpos.mx; romero.manuel@colpos.mx

F. J. A. Velasco
Universidad Autónoma de Tlaxcala, Apizaco, Mexico
e-mail: franciscojavier.albores.v@uatx.mx

J. A. Christen et al. (eds.), *Statistics, Society and Environment*, Springer Proceedings in Mathematics & Statistics 479, https://doi.org/10.1007/978-3-031-78401-9_2

means of a credible bound based on a non-informative prior. Within this Bayesian approach, the construction of CART is guided by a criterion, which combines the point estimation with the uncertainty of the estimation. This Bayesian criterion is specially useful in higher dimensions, where the uncertainty of estimations tends to increase. The aforementioned will be empirically shown by several Monte Carlo simulations based on artificially designed problems.

The continuous CART models are computed by partitioning the data space into hyper-rectangular subsets and estimating the mean answer as a local approximation in each subset. The partition is produced by splitting one variable at a time. Each splitting step is defined by two parameters: the axis to split and the splitting proportion. For any given axis, the best splitting proportion is one that minimises the squared difference between the observed data and the local approximations. Within optimisation terms, the Mean Squared Error (MSE) is considered as an objective function of the splitting proportion.

This chapter introduces a criterion function negatively proportional (NPC) to the objective function MSE. As an instance of splitting in a given axis, Fig. 1 illustrates point estimations of MSE and NPC as functions of the splitting proportion. A regular CART model is computed by minimising a point estimation of the MSE function, or equivalently, by maximising a point estimation of the NPC function. An additional method for computing a CART model is to select the splitting proportion which maximises the upper credible bound of the NPC function (See Fig. 1). This strategy combines the point estimation with the uncertainty in the prediction [3]. In this chapter, this method based on Bayesian optimisation is formally introduced by defining the criterion NPC and a bound estimator based on a non-informative prior. The use of NPC is justified by several facilities for defining a bound estimation.

The application of a Bayesian optimisation strategy to the problem of minimising MSE in CARTs is a novel approach and should be clearly distinguished from the Bayesian estimation of CART models by means of prior functions defined for trees by several authors [4–6].

Section 2 offers a formal definition of the CART model. Section 3 provides the bound estimator of the criterion NPC and an efficient numerical method

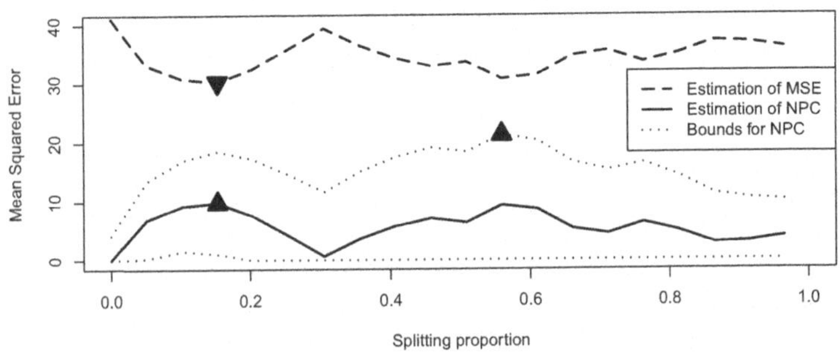

Fig. 1 Point estimations of the Mean Squared Error (MSE) and the negatively proportional criterion (NPC) as functions of the splitting proportion. A estimated bound for NPC is included to illustrate how the point estimation and the upper bound can conduce to different optima

for computing it. An experimental design based on Monte Carlo simulations is described in Sect. 4. The computational complexity of the proposed and the regular CART is analysed in Sect. 5. Finally, all results are summarised and discussed in Sect. 6 to draw conclusions.

2 Classification and Regression Trees

Classification and Regression Trees (CARTs) can be considered as a method for approximating a function f with continuous domain and codomain. In CART, as other tree-based methods [7], the feature space is partitioned into a set of hyperrectagles as described further [2].

Definition 1 A **CART tree** $\hat{f}$ is an approximation of a function $f : \mathbb{R}^n \rightarrow \mathbb{R}$ based on the local averages of f in partitions $\mathbb{G} = \{G_1, \ldots, G_t\}$ of a feasible domain $D \subset \mathbb{R}^n$. Formally,

$$\hat{f}(x) = \sum_{i=1}^{t} E\{f(X)|X \in G_i\} I(x \in G_i) \tag{1}$$

Where,

1. $\bigcup_{i=1}^{t} G_i = D$
2. $\sum_{i=1}^{n} \lambda(G_i) = \lambda(D)$, and λ stands for Lebesgue measure.
3. $X \sim \text{Uniform}(D)$

Let us consider the most simple form of partition: a bisection based on an axis threshold. Given a hyperrectagular subset $D \subset \mathbb{R}^d$, a binary partition $\mathbb{G}(j, w) = \{A(j, w), B(j, w)\}$ can be defined by means of two parameters: 1) the index of the input variable to split $j = 1, \ldots, d$; and 2) the proportion of Lebesgue measure $w \in (0, 1)$ such that $w = \lambda(A(j, w))/\lambda(D)$. The optimal bisection should minimise the mean squared error $E\{(f(X) - \hat{f}(X))^2|X \in D\}$. To determine the splitting parameters (j, w), an exhaustive search is performed by computing statistics based on an uniform sample $S = \{X_1, \ldots, X_m\} \in D$. The number of partitions defined by (j, w) that distribute different inside observations in $A(j, w)$ and $B(j, w)$ is limited by the sample size m, so an exhaustive search is feasible.

In this chapter, instead of estimating the mean squared error (MSE), a criterion negatively proportional (NPC) to MSE is introduced.

Definition 2 Let (j, w) be a pair of splitting parameters, the **negatively proportional criterion (NPC)**, denoted $C(j, w)$, is given by:

$$\begin{aligned} C(j, w) = {} & E\{f(X)|X \in A(j, w)\}^2 \Pr(X \in A(j, w)|X \in D) \\ & + E\{f(X)|X \in B(j, w)\}^2 \Pr(X \in B(j, w)|X \in D) \end{aligned}$$

Theorem 1 *Let* (j, w) *and* (i, v) *be two different pairs of splitting parameters. Then,*

$$C(j, w) > C(i, v) \Rightarrow MSE(j, w) < MSE(i, v) \tag{2}$$

Where,

$$MSE(j, w) = E\{(f(X) - \hat{f}(X))^2 | X \in A(j, w)\}^2 \Pr(X \in A(j, w) | X \in D) \\ + E\{(f(X) - \hat{f}(X))^2 | X \in B(j, w)\}^2 \Pr(X \in B(j, w) | X \in D)$$

and analogously for $MSE(i, v)$.

Proof in Appendix.

Briefly, a CART model (Definition 1) is computed by recursively splitting hyperrectangular subsets: After defining the first bisection, the largest subset in terms of Lebesgue measure is selected to be splitted, and then the process of selecting and splitting the current largest subset is repeated [8]. For the last process, there are several stop criteria [9]. As an experimental standard in this paper, the process is stopped when a maximum number of subsets is reached: maxSubsets = 10×dimensions.

3 Methods

3.1 Conservative and Optimistic Criteria

Theorem 2 states two different estimators of NPC: (1) the point estimator and (2) the upper credible bound. On the one hand, the point estimator is quite easy to compute; however, as a result of Theorem 1, the partition that maximises NPC also minimises the mean squared error. Thus, maximising the point estimator of NPC produces a regular CART model. On the other hand, within the terms of Bayesian optimisation, the upper credible bound is equivalent to an acquisition function that manages an optimistic exploration [10]. Hereafter, the point and upper credible bound estimators of NPC are called **conservative** and **optimistic** criteria respectively.

Theorem 2 *Let* $\hat{f}(x)$ *be an approximation of* $f : D \to \mathbb{R}$ *defined by means of the partition* $\mathbb{G}(j, w) = \{A(j, w), B(j, w)\}$, *and let* $S = \{x_1, x_2, \ldots, x_n\}$ *be random sample with* $X_i = x_i \sim Uniform(D)\, \forall\, i = 1, \ldots, n$. *If* $n \to \infty$, *then the negatively proportional criterion* $C(\mathbb{G}(j, w))$ *has the following estimators:*

- *Point estimator* **(Conservative criterion)**

$$\hat{C}(\mathbb{G}(j, w)) = \hat{\mu}_A^2 \pi_A + \hat{\mu}_B^2 \pi_B \tag{3}$$

– *Upper credible bound with probability* $1-\alpha$ **(Optimistic criterion)**

$$upper.\hat{C}(\mathbb{G}(j,w)) = \left[\frac{max(Z)\hat{\mu}_A}{max(Z)-\hat{\sigma}_A^2\pi_A}\right]^2\pi_A + \left[\frac{max(Z)\hat{\mu}_B}{max(Z)-\hat{\sigma}_B^2\pi_B}\right]^2\pi_B \tag{4}$$

– *Lower credible bound with probability* $1-\alpha$

$$low.\hat{C}(\mathbb{G}(j,w)) = \begin{cases}\left[\frac{min(Z)\hat{\mu}_A}{min(Z)-\hat{\sigma}_A^2\pi_A}\right]^2\pi_A + \left[\frac{min(Z)\hat{\mu}_B}{min(Z)-\hat{\sigma}_B^2\pi_B}\right]^2\pi_B & L < \chi^2_{2,1-\alpha}\\ 0 & otherwise\end{cases} \tag{5}$$

Where:

– $\pi_A = \Pr(X \in A(j,w)|X \in D) = \lambda(A(j,w))/\lambda(D)$
– $S_A \subset S = \{x|x \in A(j,w)\}$
– $n_A = |S_A| = \sum_{S_A} 1$
– $\hat{\mu}_A = \frac{1}{n_A}\sum_{x\in S_A} f(x)$
– $\hat{s}_A^2 = \frac{1}{n_A-1}\sum_{x\in S_A}[f(x)-\hat{\mu}_A]^2$
– $\hat{\sigma}_A^2 = \hat{s}_A^2/n_A$
– *Analogously for* π_B, S_B, n_B, $\hat{\mu}_B$, $\hat{s}_B^2$ *and* $\hat{\sigma}_B^2$
–

$$L = \left(\frac{\hat{\mu}_A}{\hat{\sigma}_A}\right)^2 + \left(\frac{\hat{\mu}_B}{\hat{\sigma}_B}\right)^2 \tag{6}$$

–

$$Z = \{z_\lambda \in \mathbb{R}|\hat{\sigma}_A^2\left(\frac{\hat{\mu}_A\pi_A}{\hat{\sigma}_A^2\pi_A - z_\lambda}\right)^2 + \hat{\sigma}_B^2\left(\frac{\hat{\mu}_B\pi_B}{\hat{\sigma}_B^2\pi_B - z_\lambda}\right)^2 = \chi^2_{2,1-\alpha}\} \tag{7}$$

Proof in Appendix.

3.2 Computing Estimators

At first sight, computing the set Z (Eq. 7) from the last theorem seems to be a complex task. However, the next points detail an efficient strategy to find the minimum and maximum values in Z.

1. Equation 7—that defines the set Z—can be numerically transformed into a quartic polynomial and solved by applying the power method [11] to the corresponding companion matrix. The power method computes the greatest root

in absolute value $|z_\lambda^*|$ by iterating a multiplication of the companion matrix by an initially random vector. An initial vector close to the eigenvector of the companion matrix ensures a quick convergence. This methodology computes a lower or maximal bound. The next point is useful to compute the opposite bound.

2. Let z_λ^* be a known solution of Eq. 7. Thus, $q_a^2 = (\hat{\sigma}_A\pi_A - z_\lambda^*)^2$ and $q_b^2 = (\hat{\sigma}_B\pi_B - z_\lambda^*)^2$ are known. Now, let z_λ be a unknown root, then:

$$\left.\begin{array}{l}(\hat{\sigma}_A\pi_A - z_\lambda)^2 = q_a^2 \\ (\hat{\sigma}_B\pi_B - z_\lambda)^2 = q_b^2\end{array}\right\} \Rightarrow \hat{\sigma}_A^2\left(\frac{\hat{\mu}_A\pi_A}{\hat{\sigma}_A^2\pi_A - z_\lambda}\right)^2 + \hat{\sigma}_B^2\left(\frac{\hat{\mu}_B\pi_B}{\hat{\sigma}_B^2\pi_B - z_\lambda}\right)^2 = \chi^2_{2,1-\alpha} \tag{8}$$

If and only if z_λ corresponds to a bound (See Eqs. 38, 41), then $q_a q_b > 0$:

$$z_\lambda \in (\max(\hat{\sigma}_A^2\pi_A, \hat{\sigma}_B^2\pi_B), +\infty) \Rightarrow \hat{\sigma}_A^2\pi_A - z_\lambda < 0, \hat{\sigma}_B^2\pi_B - z_\lambda < 0 \tag{9}$$
$$\Rightarrow (\hat{\sigma}_A^2\pi_A - z_\lambda)(\hat{\sigma}_B^2\pi_B - z_\lambda) > 0$$

$$z_\lambda \in (-\infty, \min(\hat{\sigma}_A^2\pi_A, \hat{\sigma}_B^2\pi_B)) \Rightarrow \hat{\sigma}_A^2\pi_A - z_\lambda > 0, \hat{\sigma}_B^2\pi_B - z_\lambda > 0 \tag{10}$$
$$\Rightarrow (\hat{\sigma}_A^2\pi_A - z_\lambda)(\hat{\sigma}_B^2\pi_B - z_\lambda) > 0$$

$$(\hat{\sigma}_A^2\pi_A - z_\lambda)(\hat{\sigma}_B^2\pi_B - z_\lambda) < 0 \Rightarrow z_\lambda \in (\min(\hat{\sigma}_A^2\pi_A, \hat{\sigma}_B^2\pi_B), \max(\hat{\sigma}_A^2\pi_A, \hat{\sigma}_B^2\pi_B)) \tag{11}$$

Therefore, the lower and upper bounds correspond to the real roots of the following quadratic equation:

$$(\hat{\sigma}_A^2\pi_A - z_\lambda)(\hat{\sigma}_B^2\pi_B - z_\lambda) = q_a q_b \tag{12}$$

3. To improve the performance of the power method, a helpful conjecture is that $\lim_{\Delta\to 0} C(\mathbb{G}(w+\Delta)) = C(\mathbb{G}(w))$. Therefore, close values of w have similar statistics and produce similar companion matrixes. For the power method, an initial vector close to the eigenvector of the companion matrix ensures a quick convergence [11]. This property is useful to compute roots with a previously computed eigenvector as the initial vector for a companion matrix produced by a close w. In consequence, for each axis, an initially random vector is modified for progressively becoming the eigenvector of different companion matrixes produced by successive splitting proportions w. This approach successfully reduces the amount of required matrix multiplications.

4 Monte Carlo Experiments

In order to consistently compare conservative and optimistic criteria, a Monte Carlo experiment is performed considering optimisation benchmark functions in several dimensions. This collection of well-known and hard-predictable functions allows to arbitrarily modify the dimension of the function, though keeping relevant characteristics such as: general shape, number of optima and a feasible domain to compute. The specific benchmark functions are listed in Table 1, whilst their corresponding formulæ and domain specification are detailed in [12].

4.1 Sign Test

An ordinary experimental design based on an ANOVA test is not possible due to heteroskedasticity. For this reason, an analysis based on multiple sign tests with Bonferroni correction is performed.

For a given benchmark function f defined in dimension $d = 10, 20, \ldots, 50$, 40 Monte Carlo comparisons between conservative and optimistic criteria are performed. Each Monte Carlo run $t = 1, \ldots, 40$ consists of computing a Bernoulli random variable B_t by the following steps:

1. Draw an uniform sample S over the feasible domain D. Formally, $S = \{S_1 = s_1, \ldots, S_n = s_n\}$ where $S_i \sim Uniform(D)\, \forall i = 1, \ldots, n$. The sample size is defined according to dimension $n = 1500\text{d}$.
2. Divide the sample S into two categories: training sample $S_{\text{train}} = \{s_1, \ldots, s_{n/2}\}$ and testing sample $S_{\text{test}} = \{s_{n/2+1}, \ldots, s_n\}$.
3. Compute a CART model $\hat{f}_{\text{cons}}$ by using the conservative splitting criterion (point estimation of C) and the training sample S_{train}.
4. Compute a CART model $\hat{f}_{\text{opti}}$ by using the optimistic splitting criterion (upper bound estimation of C) and the same training sample S_{train}.
5. Estimate the mean squared error of both models $\hat{f}_{\text{cons}}$ and $\hat{f}_{\text{opti}}$ by using the same testing sample S_{test}:

$$MSE_{\text{cons}} = \frac{1}{|S_{\text{test}}|} \sum_{s \in S_{\text{test}}} [f(s) - \hat{f}_{\text{cons}}(s)]^2 \tag{13}$$

$$MSE_{\text{opti}} = \frac{1}{|S_{\text{test}}|} \sum_{s \in S_{\text{test}}} [f(s) - \hat{f}_{\text{opti}}(s)]^2 \tag{14}$$

6. If $MSE_{\text{opti}} < MSE_{\text{cons}}$, i.e. the optimistic criterion is better in terms of mean squared error, then the Bernoulli variable counts a success $B_t = 1$. Otherwise, i.e. the conservative criterion is better than or equal to the optimistic criterion, $B_t = 0$.

Table 1 Number of successes (the optimistic criterion is better than the conservative one) after 40 trials; each row corresponds to the same function defined in several dimensions

Function	Dimension 10	20	30	40	50
Ackley	8.00	19.00	25.00	40.00	40.00
Griewank	21.00	5.00	37.00	40.00	40.00
Levy	2.00	13.00	19.00	22.00	35.00
Michalewicz	18.00	23.00	40.00	40.00	40.00
Rastrigin	17.00	30.00	28.00	40.00	40.00
Schwefel	13.00	27.00	36.00	34.00	37.00
Sphere	4.00	33.00	32.00	35.00	33.00
Zakharov	0.00	8.00	16.00	35.00	32.00

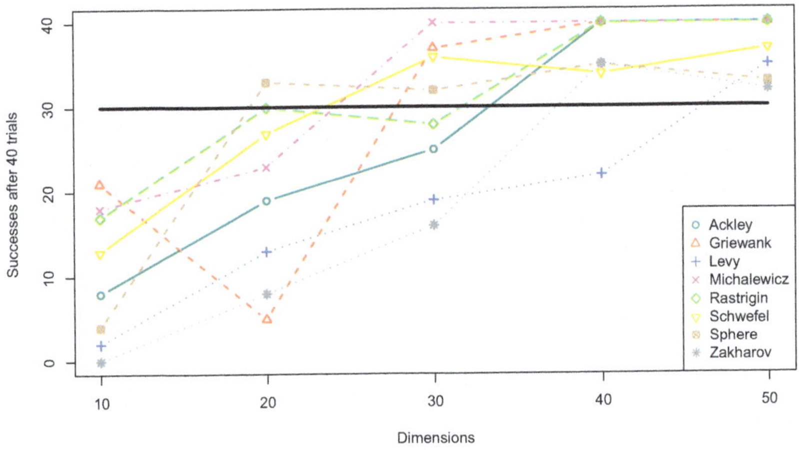

Fig. 2 Evolution of the number of successes according to the dimension for each benchmark function

Therefore, the sum $A = \sum_{t=1}^{40} B_t$ is a Binomial random variable defined by:

$$A \sim Binomial(40, p) \quad \text{where } p = \Pr(MSE_{\text{opti}} < MSE_{\text{cons}}) \tag{15}$$

Under the null hypothesis $H_0 : p \leq 0.5$ (i.e. the conservative criterion is better than or equal to the optimistic criterion), consider that $\Pr(30 \leq A \leq 40) \leq 0.0011$. Thus, to contrast the null hypothesis for eight benchmark functions, the interval $[30, 40]$ is a rejection region for H_0 with $\alpha < 0.01/8$ (Bonferroni correction) in favour of $H_1 : p > 0.5$, i.e. the optimistic criterion is better.

Table 1 contains the corresponding statistic A for eight different benchmark functions, each one of them defined in dimension $d = 10, 20, \ldots, 50$. Figure 2 illustrates Table 1 as a collection of series that show the evolution of the number of successes according to the dimension for each benchmark function. An horizontal line in the plot delimits the rejection region $[30, 40]$. In Fig. 2, each series represents a different function. If a segment of a series is over the horizontal bound, the null hypothesis—the conservative criterion is better or equal to the optimistic criterion—is rejected.

In dimensions $d = 20$, the null hypothesis is rejected only for two functions: Grienwank and Sphere. In dimension $d = 30$, the null hypothesis is rejected for the half of tested functions. Finally, in dimensions $d = 40, 50$, the null hypothesis is rejected for all functions, except for Levy in dimension 40. Furthermore, all series show positive trend, i.e. the number of successes increases in higher dimensions. *Therefore, in dimensions* > 30, *an optimistic criterion decreases the mean squared error based on observations not included in the training set.*

4.2 Relative Efficiency

The regression trees f_{opti} (the proposal) and f_{cons} (the regular CART) constitute estimators of a given function f. These estimators can be compared by means of the relative efficiency [13], which is defined as the ratio of the variances, or equivalently by

$$\text{eff}(\hat{f}_{\text{cons}}, \hat{f}_{\text{opti}}) = \frac{Var\{\hat{f}_{\text{cons}}\}}{Var\{\hat{f}_{\text{opti}}\}} = \frac{E\{[\hat{f}_{\text{cons}}(X) - f(X)]^2\}}{E\{[\hat{f}_{\text{opti}}(X) - f(X)]^2\}} = \frac{E\{MSE_{\text{cons}}\}}{E\{MSE_{\text{opti}}\}} \tag{16}$$

If $\text{eff}(\hat{f}_{\text{cons}}, \hat{f}_{\text{opti}}) > 1$, then the estimator $\hat{f}_{\text{opti}}$ is preferred over $\hat{f}_{\text{cons}}$. When comparing regression trees, the ratio eff depends on the given objective function f, so eff has no general closed-form expression. Some authors [14] suggest a simulation-based approach, which consists of computing the empirical distribution of the ratio eff by iteratively estimating pairs of the models to be compared. With this approach, for each different function f and dimension d, Table 2 contains the approximation of $\Pr\left(\frac{MSE_{\text{cons}}}{MSE_{\text{opti}}} > 1\right)$ and $E\left\{\frac{MSE_{\text{cons}}}{MSE_{\text{opti}}}\right\}$ based on the empirical distribution of the ratios observed in the previous Monte Carlo experiment.

Consider that $\Pr\left(\frac{MSE_{\text{cons}}}{MSE_{\text{opti}}} > 1\right) = \Pr(MSE_{\text{cons}} > MSE_{\text{opti}}) \approx A/40$, where A is the corresponding Binomial variable (Eq. 15) contained in Table 1. In consequence, the left-side probabilities in Table 2 are redundant to the sign test based on the same Monte Carlo experiment. To avoid drastic probabilities such as 0 or 1, a Bayesian estimation is performed under the following assumption of normality with known mean and precision with prior gamma:

$$f_{\text{opti}}(x) \sim \mathcal{N}(f(x), 1/\tau_{\text{opti}}) \; \forall \, x \in D$$

$$f_{\text{cons}}(x) \sim \mathcal{N}(f(x), 1/\tau_{\text{cons}})$$

$$\tau_{\text{opti}}, \tau_{\text{cons}} \sim Gamma(\alpha_0, \beta_0)$$

Table 2 Approximations about $\frac{MSE_{\text{cons}}}{MSE_{\text{opti}}}$ based on the empirical density

	$\Pr\left(\frac{MSE_{\text{cons}}}{MSE_{\text{opti}}} > 1\right)$					$E\left\{\frac{MSE_{\text{cons}}}{MSE_{\text{opti}}}\right\}$				
Function, $d =$	10	20	30	40	50	10	20	30	40	50
Ackley	0.2000	0.4750	0.6250	1.0000	1.0000	0.7472	0.9943	1.1021	1.1472	1.2800
Griewank	0.5250	0.1250	0.9250	1.0000	1.0000	1.1156	0.7226	1.2805	1.5234	1.7782
Levy	0.0500	0.3250	0.4750	0.5500	0.8750	0.5747	0.9004	0.9861	1.0883	1.1223
Michalewicz	0.4500	0.5750	1.0000	1.0000	1.0000	0.9494	1.0290	1.2220	1.4120	1.3314
Rastrigin	0.4250	0.7500	0.7000	1.0000	1.0000	0.9073	1.0694	1.1052	1.2376	1.2973
Schwefel	0.3250	0.6750	0.9000	0.8500	0.9250	0.9027	1.1082	1.1402	1.1074	1.2136
Sphere	0.1000	0.8250	0.8000	0.8750	0.8250	0.7395	1.1356	1.1369	1.1925	1.2052
Zakharov	0.0000	0.2000	0.4000	0.8750	0.8000	0.3152	0.8078	0.9428	1.0766	1.0639

If the prior parameters $\alpha_0, \beta_0 \rightarrow 0$, then the Gamma distribution becomes a non-informative prior [15]. The posterior distributions of τ_{opti}, τ_{cons} and $\tau_{\text{opti}}/\tau_{\text{cons}}$ are conditionally defined by the set D of observations computed for the sign test exposed in the previous subsection.

$$\tau_{\text{opti}}|D \sim Gamma\left(\alpha_{\text{opti}} = 40/2, \beta_{\text{opti}} = \frac{1}{2}\sum_{t=1}^{40} MSE_{\text{opti}}[t]\right)$$

$$\tau_{\text{cons}}|D \sim Gamma\left(\alpha_{\text{cons}} = 40/2, \beta_{\text{cons}} = \frac{1}{2}\sum_{t=1}^{40} MSE_{\text{cons}}[t]\right)$$

$$\tau_{\text{opti}}/\tau_{\text{cons}}|D \sim B'(\alpha_{\text{cons}}, \alpha_{\text{opti}}, \beta_{\text{cons}}/\beta_{\text{opti}})$$

Thus, the ratio $MSE_{\text{cons}}/MSE_{\text{opti}} = \tau_{\text{opti}}/\tau_{\text{cons}}$ has a Beta prime distribution. Table 3 contains the approximation of $\Pr\left(\frac{MSE_{\text{cons}}}{MSE_{\text{opti}}} > 1|D\right)$, $E\left\{\frac{MSE_{\text{cons}}}{MSE_{\text{opti}}}|D\right\}$ based on the posterior Beta prime distribution. Although mutually consistent, the probabilities approximated in Tables 2 and 3 may subtly differ. However, the expected values of the ratio eff from both tables are quite close to each other.

The ratios from Tables 2 and 3 confirm that the proposed optimistic estimation is better in dimensions $d > 30$. Remarkably, in the lowest dimensions tested ($d \leq 20$), there exist objective functions for which the optimistic estimation is worse than the conservative. Even for these least favourable cases, the optimistic estimation becomes better than the conservative as dimension increases.

Table 3 Approximations about $\frac{MSE_{\text{cons}}}{MSE_{\text{opti}}}$ based on Bayesian estimation

	$\Pr\left(\frac{MSE_{\text{cons}}}{MSE_{\text{opti}}} > 1 \mid D\right)$					$E\left\{\frac{MSE_{\text{cons}}}{MSE_{\text{opti}}} \mid D\right\}$				
Function, $d =$	10	20	30	40	50	10	20	30	40	50
Ackley	0.2265	0.4879	0.6865	0.9550	0.9959	0.7502	0.9955	1.1033	1.1471	1.2803
Griewank	0.6137	0.1195	0.8942	0.9946	0.9987	1.1273	0.7237	1.2817	1.5237	1.7801
Levy	0.1028	0.3187	0.4770	0.6535	0.7633	0.5756	0.9023	0.9882	1.0903	1.1233
Michalewicz	0.4315	0.5469	0.9800	0.9700	0.9973	0.9513	1.0337	1.2224	1.4118	1.3312
Rastrigin	0.3452	0.5807	0.7189	0.9952	0.9964	0.9065	1.0698	1.1061	1.2375	1.2975
Schwefel	0.3683	0.7125	0.8417	0.7649	0.9148	0.9056	1.1085	1.1409	1.1080	1.2153
Sphere	0.1314	0.8315	0.7457	0.9634	0.9052	0.7425	1.1366	1.1385	1.1930	1.2055
Zakharov	0.0023	0.1178	0.4181	0.7099	0.6029	0.3165	0.8081	0.9453	1.0770	1.0661

5 Computational Performance

In contrast to relative efficiency, the computational cost associated to each estimator can be expressed in a closed form. Both models *opti* and *cons* are computed with the same algorithm, except in the step for defining a partition.

1. **Conservative estimation.** The computation time t_{cons} for a partition is a function of the sample size n and the number of dimensions d of f. For each dimension d, there exist $n-1$ different partitions which separate an n-size sample into different subsets A and B. To identify the best partition, statistics of each pair of subsets A and B are computed. Therefore,

$$t_{\text{cons}}(d, n) \propto d(n-1)n \tag{17}$$

2. **Optimistic estimation.** After computing statistics for every subset A and B from $d(n-1)$ different partitions, the set Z (Eq. 7) must be defined for each partition. The set Z is found using the power method, which consists of iterating the multiplication of a 4×4 matrix. The dimensions of the last matrix do not depend on n or d, so this step can be associated with a constant $T_{4\times4}$. Therefore,

$$t_{\text{opti}}(d, n) \propto d(n-1)n + d(n-1)T_{4\times4} \tag{18}$$

It is easy to prove that $t_{\text{opti}}(d, n) > t_{\text{cons}}(d, n)$, i.e. the proposed optimistic estimation is slower than conservative estimation (regular CART). However, this difference is left out in the big $\mathbb{O}$ notation,

$$\mathbb{O}(t_{\text{cons}}(d, n)) = \mathbb{O}(t_{\text{opti}}(d, n)) = n^2 \tag{19}$$

Consequently, both estimators have the same *polynomial time* limit. According to the analysis principles exposed in [16], their time difference is not drastically significant, and both remain equally feasible.

6 Discussion and Conclusion

In this chapter, a Bayesian optimisation technique is applied to improve the estimation of a continuous CART model. The proposed tree f_{opti} and the regular CART f_{cons} have been compared through statistics based on a Monte Carlo simulation. The following are remarkable points to contrast both models:

1. **Sign test.** (Section 4.1) A sign test was performed to compare the error of both models measured with observations not included in the training set. In dimensions >30, the error of the proposed CART is statistically lower than the error of the regular CART. Therefore, the proposal may be an excellent alternative for increasing the validation rate in higher dimensions.
2. **Estimation of the relative efficiency.** (Section 4.2) In the lowest dimensions tested ($d \leq 20$), there exist objective functions f for which the proposed estimation f_{opti} is worse than the regular f_{cons}. However, even for these initially least favourable functions, the proposed estimation becomes better than the regular as dimension increases.
3. **Computational cost.** (Section 5) Both models have the same *polynomial time* limit, so both are equally feasible.

The point 2 states that the proposed CART estimation may underperform in some cases. This behaviour is supported by the following theoretical nuances:

- **No-Free Lunch Theorem.** The no-free lunch theorem enounces the impossibility of concluding that some algorithm is better than other across all scenarios [17]. Thus, the relative performance of the proposal (as any other algorithm) depends on the specific problem (function and dimension).
- **Performance of an optimistic acquisition function.** In Bayesian optimisation terms, the proposed estimation based on the upper credible bound constitutes an exploration based on an optimistic acquisition function [18]. Some authors suggest that an optimistic acquisition function is efficient for exploring promising solutions mostly in higher dimensions [3, 18]. This behaviour could be originated by the increasing variance of the point estimators in high-dimensional functions. Consider that, for all functions, the efficiency of the proposal consistently keeps a positive trend as dimension increases.

In this regard, the authors suggest that this version of CART be applied and tested alongside a regular CART model. A thoughtful evaluation considering the relative efficiency for the specific problem is essential for real-world applications. Some scenarios where the proposal may be extended or directly applied are:

1. **Data sets with continuous and categorical variables.** The original CART models can be estimated for data spaces with continuous and categorical variables. For each variable, the splitting process is independent. Thus, if the selected variable to split is continuous, the Bayesian proposal may be applied. If the variable is categorical, then the standard CART splitting criterion must be

applied. In light of these considerations, as future work, it would be beneficial to evaluate the application of this proposal to real-world datasets with even more than 50 continuous variables.

2. **Search based on trees.** In the context of numerical optimisation, some authors propose the use of tree-models for performing a search [19, 20]. The proposed CART has been empirically validated with tests based on well-known optimisation benchmark functions, so a potential application is to guide the exploration of an optimisation process.

In summary, the proposed model shows statistically lower error in higher dimensions (>30) compared to the regular CART, but it might underperform in lower dimensions. This is attributed to the nature of the optimistic Bayesian approach and the No-Free Lunch Theorem, which implies that algorithms perform differently based on data and dimensions. The promising performance of this Bayesian approach motivates that this version of CART be tested alongside the regular CART to evaluate which model is better for specific real-world problems.

Acknowledgments The authors wish to thank two anonymous referees for their really valuable comments and recommendations, which have significantly improved the chapter.

Disclosure of Interests The authors have no competing interests to declare.

Proofs

Proof of Theorem 1 As a preliminary step, the following statement is demonstrated. For simplicity, consider $A = A(j, w)$ and $B = B(j, w)$.

$$E\{f(X)^2|X \in D\} - C(j, w) = \int_D f(x)^2 \frac{dx}{\lambda(D)} - C(j, w) \tag{20}$$

$$= \int_D f(x)^2 \frac{dx}{\lambda(D)} - \left(\int_A f(x)\frac{dx}{\lambda(A)}\right)^2 \frac{\lambda(A)}{\lambda(D)} - \left(\int_B f(x)\frac{dx}{\lambda(B)}\right)^2 \frac{\lambda(B)}{\lambda(D)} \tag{21}$$

$$= \left[\int_A f(x)^2 \frac{dx}{\lambda(A)} - \left(\int_A f(x)\frac{dx}{\lambda(A)}\right)^2\right] \frac{\lambda(A)}{\lambda(D)} + \left[\int_B f(x)^2 \frac{dx}{\lambda(B)} - \left(\int_B f(x)\frac{dx}{\lambda(B)}\right)^2\right] \frac{\lambda(B)}{\lambda(D)} \tag{22}$$

$$= Var\{f(X)|X \in A(j, w)\} \Pr(X \in A(j, w)|X \in D)$$
$$+ Var\{f(X)|X \in B(j, w)\} \Pr(X \in B(j, w)|X \in D) \tag{23}$$

$$= MSE(j, w) \tag{24}$$

Now, from the initial inequality,

$$C(j, w) > C(i, v) \tag{25}$$

$$-C(j, w) < -C(i, v) \tag{26}$$

$$E\{f(X)^2|X \in D\} - C(j, w) < E\{f(X)^2|X \in D\} - C(i, v) \tag{27}$$

$$MSE(j, w) < MSE(i, v) \tag{28}$$

□

Proof of Theorem 2 This proof consists of finding the minimum and maximum criteria C subject to parameters enclosed by a credible region.

1. Consider $Z_A = E\{f(X)|X \in A(w)\}$ and $Z_B = E\{f(X)|X \in B(w)\}$ with $X \sim Uniform(D)$. Then, by Jeffreys' prior, or equivalently, by Central Limit Theorem:

$$Z_A = z_A|S \sim \text{Normal}(\hat{\mu}_A, \hat{\sigma}_A), \text{ or well, } \lim_{n\to\infty} \frac{\hat{\mu}_A - Z_A}{\hat{\sigma}_A} \sim \text{Normal}(0, 1) \tag{29}$$

$$Z_B = z_B|S \sim \text{Normal}(\hat{\mu}_B, \hat{\sigma}_B), \text{ or well, } \lim_{n\to\infty} \frac{\hat{\mu}_B - Z_B}{\hat{\sigma}_B} \sim \text{Normal}(0, 1) \tag{30}$$

In both cases, at $100(1-\alpha)\%$ of times, the actual population means are contained by the following ellipse:

$$\left(\frac{\hat{\mu}_A - z_A}{\hat{\sigma}_A}\right)^2 + \left(\frac{\hat{\mu}_B - z_B}{\hat{\sigma}_B}\right)^2 \leq \chi^2_{2,1-\alpha} \tag{31}$$

The lower and upper bounds of the criterion NPC are found by the method of Lagrange multipliers.

- **Optimise:** $z_A^2\pi_A + z_B^2\pi_B$ (See Definition 2)
- **Subject to:** $\left(\frac{\hat{\mu}_A - z_A}{\hat{\sigma}_A}\right)^2 + \left(\frac{\hat{\mu}_B - z_B}{\hat{\sigma}_B}\right)^2 = \chi^2_{2,1-\alpha}$
- **Lagrangian function:**

$$L(z_a, z_b, z_\lambda) = z_A^2\pi_A + z_B^2\pi_B - z_\lambda \left[\left(\frac{\hat{\mu}_A - z_A}{\hat{\sigma}_A} \right)^2 + \left(\frac{\hat{\mu}_B - z_B}{\hat{\sigma}_B} \right)^2 - \chi^2_{2,1-\alpha} \right] \quad (32)$$

$$\nabla L(z_A, z_B, z_\lambda) = \mathbf{0} \Rightarrow \begin{cases} z_A = \frac{z_\lambda \hat{\mu}_A}{z_\lambda - \hat{\sigma}_A^2 \pi_A} \\ z_B = \frac{z_\lambda \hat{\mu}_B}{z_\lambda - \hat{\sigma}_B^2 \pi_B} \\ \hat{\sigma}_A^2 \left(\frac{\hat{\mu}_A \pi_A}{\hat{\sigma}_A^2 \pi_A - z_\lambda} \right)^2 + \hat{\sigma}_B^2 \left(\frac{\hat{\mu}_B \pi_B}{\hat{\sigma}_B^2 \pi_B - z_\lambda} \right)^2 = \chi^2_{2,1-\alpha} \end{cases} \quad (33)$$

$$\det H(L) = \frac{8}{\hat{\sigma}_A^2}(z_\lambda - \pi_A \hat{\sigma}_A^2) \left[\frac{\mu_B \pi_B}{\hat{\sigma}_B^2 \pi_B - z_\lambda} \right]^2 + \frac{8}{\hat{\sigma}_B^2}(z_\lambda - \pi_B \hat{\sigma}_B^2) \left[\frac{\mu_A \pi_A}{\hat{\sigma}_A^2 \pi_A - z_\lambda} \right]^2 \quad (34)$$

2. Instead of finding solutions of the system of Eqs. 33, some intervals which contain solutions are defined. To this end, consider the following function:

$$K(z_\lambda) = \hat{\sigma}_A^2 \left(\frac{\hat{\mu}_A \pi_A}{\hat{\sigma}_A^2 \pi_A - z_\lambda} \right)^2 + \hat{\sigma}_B^2 \left(\frac{\hat{\mu}_B \pi_B}{\hat{\sigma}_B^2 \pi_B - z_\lambda} \right)^2 - \chi^2_{2,1-\alpha} \quad (35)$$

- $\lim_{z_\lambda \to +\infty} K(z_\lambda) = -\chi^2_{2,1-\alpha}$ is negative.
- $\lim_{z_\lambda \to -\infty} K(z_\lambda) = -\chi^2_{2,1-\alpha}$ is negative.
- $\lim_{z_\lambda \to \sigma_A^2 \pi_A} K(z_\lambda) = +\infty$ is positive.
- $\lim_{z_\lambda \to \sigma_B^2 \pi_B} K(z_\lambda) = +\infty$ is positive.

It is possible to define two open intervals where $K(z_\lambda)$ is a continuous function. Each of the following intervals contains at least a real solution z_λ^* of the system of Eqs. 33 by the intermediate value theorem:

$$z_\lambda^* \in (\max(\sigma_A^2\pi_A, \sigma_B^2\pi_B), +\infty) \Rightarrow \max(\sigma_A^2\pi_A, \sigma_B^2\pi_B) < z_\lambda^* \quad (36)$$

$$\Rightarrow \det H(L(z_\lambda^*)) > 0 \quad (37)$$

$$\Rightarrow z_\lambda^* \text{ corresponds to a upper bound} \quad (38)$$

$$z_\lambda^* \in (-\infty, \min(\sigma_A^2\pi_A, \sigma_B^2\pi_B)) \Rightarrow z_\lambda^* < \min(\sigma_A^2\pi_A, \sigma_B^2\pi_B) \quad (39)$$

$$\Rightarrow \det H(L(z_\lambda^*)) < 0 \quad (40)$$

$$\Rightarrow z_\lambda^* \text{ corresponds to a lower bound} \quad (41)$$

In this regard, the greatest root of the Eq. 35 corresponds to the upper bound, and similarly, the lowest root corresponds to the lower bound. The last sentence is mathematically enounced in Eq. 5,4,7.

3. The area of the ellipse (from Eq. 33) can be reduced to zero to calculate a point estimation instead of a credible interval.

$$\lim_{z_\lambda \to \infty} \hat{\sigma}_A^2 \left(\frac{\hat{\mu}_A \pi_A}{\hat{\sigma}_A^2 \pi_A - z_\lambda} \right)^2 + \hat{\sigma}_B^2 \left(\frac{\hat{\mu}_B \pi_B}{\hat{\sigma}_B^2 \pi_B - z_\lambda} \right)^2 = 0 \tag{42}$$

$$\Rightarrow \begin{cases} z_A = \lim_{z_\lambda \to \infty} \frac{z_\lambda \hat{\mu}_A}{z_\lambda - \hat{\sigma}_A^2 \pi_A} = \hat{\mu}_A \\ z_B = \lim_{z_\lambda \to \infty} \frac{z_\lambda \hat{\mu}_B}{z_\lambda - \hat{\sigma}_B^2 \pi_B} = \hat{\mu}_B \end{cases} \Rightarrow \hat{C}(\mathbb{G}(j, w)) = \hat{\mu}_A^2 \pi_A + \hat{\mu}_B^2 \pi_B \tag{43}$$

Another method to prove the last estimator is by means of the functional equivariance property [21] of the maximum likelihood estimators $\hat{\mu}_A$ and $\hat{\mu}_A$.

4. **Correction for the lower bound.** If the cartesian origin $(0, 0)$ is contained by the ellipse (Eq. 31), then the actual population means could take the values $E\{f(X)|X \in A(w)\} = 0$ and $E\{f(X)|X \in B(w)\} = 0$. In that case, the criterion C takes the lowest possible value: $C(\mathbb{G}(w)) = 0$. Therefore,

$$low.\hat{C}(\mathbb{G}(j, w)) = \begin{cases} \left[\frac{min(Z)\hat{\mu}_A}{min(Z) - \hat{\sigma}_A^2 \pi_A} \right]^2 \pi_A + \left[\frac{min(Z)\hat{\mu}_B}{min(Z) - \hat{\sigma}_B^2 \pi_B} \right]^2 \pi_B & L < \chi^2_{2,1-\alpha} \\ 0 & \text{otherwise} \end{cases} \tag{44}$$

□

References

1. Steinberg, D.: Cart: classification and regression trees. In: The top ten algorithms in data mining, pp. 193–216. Chapman and Hall/CRC, London (2009)
2. Hastie, T., et al.: The Elements of Statistical Learning. Springer New York, New York, NY (2009), https://doi.org/10.1007/978-0-387-84858-7
3. De Ath, G.e.: Greed is good: Exploration and exploitation trade-offs in bayesian optimisation. ACM Trans. Evol. Learn. Optim. 1(1) (apr 2021), https://doi.org/10.1145/3425501
4. Chipman, H.A., et al.: Bayesian cart model search. Journal of the American Statistical Association 93(443), 935–948 (1998), https://doi.org/10.1080/01621459.1998.10473750
5. Wu, Y., Tjelmeland, H., West, M.: Bayesian cart: Prior specification and posterior simulation. Journal of Computational and Graphical Statistics 16(1), 44–66 (2007), https://doi.org/10.1198/106186007X180426
6. Rocková, V., et al.: Posterior concentration for bayesian regression trees and forests. Annals of Statistics 48(4), 2108–2131 (2020), https://doi.org/10.1214/19-AOS1879

7. Loh, W.Y.: Classification and regression trees. Data mining and knowledge discovery 1(1), 14–23 (2011), https://doi.org/10.1002/widm.8
8. Breiman, L., et al: Classification and Regression Trees. Chapman and Hall/CRC (1984), https://doi.org/10.1201/9781315139470
9. Loh, W.Y.: Fifty years of classification and regression trees. International Statistical Review 82(3), 329–348 (2014), https://doi.org/10.1111/insr.12016
10. Archetti, F., Candelieri, A.: The Acquisition Function, pp. 57–72. Springer International Publishing, Cham (2019), https://doi.org/10.1007/s10898-019-00860-4
11. Bolch, G., Greiner, S., De Meer, H., Trivedi, K.S.: Queueing networks and Markov chains. John Wiley & Sons, New Jersey (2006)
12. Thomas Back: Evolutionary algorithms in theory and practice. Oxford University Press, New York, NY (Jan 1996)
13. Wackerly, D., Owen, W., Mendenhall, W., Scheaffer, R.: Mathematical Statistics with Applications, 7th. Brooks/Cole Cengage Learning (2007)
14. Hayes, T., Usami, S., Jacobucci, R., McArdle, J.J.: Using classification and regression trees (cart) and random forests to analyze attrition. Psychol Aging 30(4), 911–929 (Dec 2015), https://doi.org/10.1037/pag0000046
15. Gelman, A.: Prior distributions for variance parameters in hierarchical models. Bayesian Analysis 1(3), 515–533 (2006), https://doi.org/10.1214/06-BA117A
16. Goldreich, O.: Computational complexity: a conceptual perspective. Cambridge University Press (2008), https://doi.org/10.1017/CBO9780511804106
17. Sterkenburg, T.F., Grünwald, P.D.: The no-free-lunch theorems of supervised learning. Synthese 199(3–4), 9979–10015 (2021), https://doi.org/10.1007/s11229-021-03233-1
18. Garnett, R.: Bayesian Optimization. Cambridge University Press (2023)
19. Hutter, F., et al.: Sequential model-based optimization for general algorithm configuration. In: Learning and Intelligent Optim. LION 2011 LNCS. vol. 6683. Springer, Berlin, Heidelberg (2011), https://doi.org/10.1007/978-3-642-25566-3_40
20. Jones, D.: The direct algorithm: 25 years later. J Glob Optim 79, 521–566 (2021), https://doi.org/10.1007/s10898-020-00952-6
21. Casella, G., Berger, R.L.: Statistical inference. Cengage Learning, Pacific Grove, CA (2021)

A Little Bird Told Me: Blueprint Toward Frequent and Representative Official Statistical Information, Through Joint Use of Social Network Posts and Survey Data

Alfredo Bustos, Silvia Fraustro, Noemí López, and Ricardo Olvera

Abstract A proposal is introduced for National Statistical Offices (NSOs) to produce representative information, on multiple topics, more frequently, jointly using household survey data and social network (SN) posts. SN posts by respondents are tagged using survey responses to train machine learning (ML) algorithms. Trained algorithms tag large volumes of recent posts. Those by the same author are tallied to tag users and form a non-random large sample for statistical production. Further monitoring is carried out by tagging posts dated between survey rounds from new "large samples." To deal with selection bias in user populations, sociodemographic (SD) variables collected during the survey are used to develop SD-tagged large sample databases of recent authors. They will be SD post-stratified during thematic studies to mitigate selection bias. Suggestions to link user-respondent survey responses and network posts are explored. An author can be tagged according to many surveys, leading to a multi-topic large-sample, to explore interactions rarely studied.

Keywords Official statistics · Machine learning · Social networks · Tagging · Representativeness · Selection bias

1 Introduction

The possibility of complementing traditional data sources, by incorporating the so-called Big Data, has produced extensive bibliography in recent years. Numerous national and international organizations have carried out various actions to study the introduction of this data source in their daily activities. The Dutch Centraal Bureau voor de Statistiek (CBS) is among the first national statistical offices to

A. Bustos (✉) · S. Fraustro · N. López · R. Olvera
Instituto Nacional de Estadística y Geografía, Aguascalientes, Mexico
e-mail: alfredo.bustos@inegi.org.mx

J. A. Christen et al. (eds.), *Statistics, Society and Environment*, Springer Proceedings in Mathematics & Statistics 479, https://doi.org/10.1007/978-3-031-78401-9_3

Table 1 GWG task forces

1. Social media data	2. Training, skills, and capacity building
3. Mobile phone data	4. Improve access to BIG Data sources
5. Cross-cutting issues	6. Promotion and communication
7. Big Data and the SDGs	8. Satellite images and geospatial data

initiate the study of these alternative sources, as well as their implications for official statistics (see [13] and [14]). In turn, the United Nations (UN) created the Global Working Group (GWG) for Big Data in Official Statistics in 2014 (see [7, 10, 18]). According to [11], its terms of reference assign the GWG, among others, the tasks of "providing a strategic vision, direction, and coordination of a global Big Data program for official statistics, including indicators for the 2030 agenda, of sustainable development. It also promotes the practical use of Big Data sources, the promotion of capacity development, training, and the exchange of experiences." During its first meeting in October 2014 in Beijing, the GWG established eight task forces enlisted in Table 1. The working groups have met and reached different progress. Jansen [7] provided a summary for some groups created by the GWG. It should be noted that no progress is reported for the teams on data integration, nor that which refers to the use of information from social networks, which is declared in recess. An immediate scope for the GWG's work is given by the 2030 Agenda for Sustainable Development, which adopted a comprehensive global monitoring framework encompassing 231 indicators within 17 goals. (See [17]). As [9] point out, "leveraging new and existing data sources (both from the public and private sectors) in order to monitor progress towards the SDGs, as well as to achieve it, is not without challenges." They emphasize that the difference in what they call "datafication" between developed and developing economies will prevent the latter from using optimally the information available. The same authors point out that "it is important to remember that despite the large body of literature and applications that already exist, the state of the art in applications focused on the innovative development of these new data sources is still in its embryonic stages." Additionally, they point out that it will be necessary to pay particular attention to "address the ethical and privacy dilemmas" that will arise from these data sources.Little has been written about the so-called selection bias in social networks (see [5] or [15]) and even less about its implications in official statistics. Iacus et al. [6] develop a methodological proposal to correct said bias, through commonly used models for estimation in small domains. In this case, the models are adjusted to those aggregated at the domain level, affected by selection bias, using similar aggregates from official statistics as explanatory variables. In other words, the aim is to correct aggregates instead of the database itself. Something similar is attempted by Tam et al. [16]. For Latin America, [3] "highlights the risks and opportunities that Big Data presents to national statistical offices in Latin America in the context of the SDGs." Risks include institutional barriers to managing change and innovation, restrictions on access and completeness of data, technical challenges, gaps in human capabilities, methodological challenges, and ethical and political risks. On the other hand, five trends stand out that they consider propitious in the region: the Latin American experience in the open data movement; the appearance of public and private associations regarding Big Data; the presence of strong committees, institutions, and working groups that encompass the entire region; the development of adaptable best practices; and the existence of an interdisciplinary innovation network that involves both national statistical offices and other actors. They develop a regional roadmap for the use of Big Data in official statistics and for monitoring the SDGs. Even considering legislation to allow ONS access to Big Data sources, as suggested in [4], it cannot be guaranteed that social media companies will continue to function. There are experiences of social networks that have seen their membership reduced or even disappeared; e.g., Sixdegrees, Friendster, My Space. The production of official statistics will have to establish with adequate anticipation the strategies to be implemented to guarantee the continuity and comparability of the results.

1.1 Mexican Twitter-Users' Mood

The project Mexican Twitter-users' mood[1] is the best-known use INEGI has given to automated learning, as well as texts from Twitter. At its inception, it was unclear whether users of this network who published georeferenced posts represented only themselves or could be used to draw inferences for the Mexican population altogether. Hence the name of the project. The procedure followed at the time may be summarized as follows. A random sample was obtained from captured messages. Each post was tagged manually by one or more students. They assessed the author's sentiment (positive, neutral, or negative) at the time of publication. This resulted in a set of tagged tweets split to form training and testing sets. Various algorithms were trained and evaluated with this input to choose suitable ones. As a result, it became possible to automatically tag millions of messages stored in our database, as well as those captured later, to monitor the mood of Twitter users daily. The uncertainty behind the title of the project "Mexican Twitter users' mood" motivated the inclusion of new items on the use of social networks in the National Survey of Availability and Use of Information Technologies in Households (ENDUTIH, Spanish acronym). Based on this information, sociodemographic discrepancies between active network and Mexican populations could be measured (see [2]). The authors highlight that for the Mexican case, the populations of social network users are, on average, younger, with higher socioeconomic levels, and higher educational status than the rest of the population. They also point out some gender biases that become apparent for some age groups.

[1] At https://www.inegi.org.mx/app/animotuitero/#/app/multiline. Post collection is not taking place anymore, given policy change by X (formerly Twitter).

1.2 Limitations

Despite the aforementioned, it has not been possible to implement a strategy to approach the desired "The Mood of Mexicans, according to Twitter." This would require sociodemographic (SD) tagging of captured tweets. However, since ENDUTIH has people, not tweets, as units of analysis, and since we have characterized selection bias and its possible correction in terms of the former, it is desirable to learn how to turn information on posts into information on their authors. If SD characteristics of Twitter users were known, at the time of calculating totals for the construction of indices, their thematic labels could be re-weighted according to the relationship between the relative sizes of the class to which they belong in the open population, on the one hand, and in that of Twitter users who published during the period in question, on the other. For the class identified with the labels *i,j,…,k*, whose relative sizes to the reference population (R) are represented by $\%P^{R}_{ij\ldots k}$ and by $\%P^{SN}_{ij\ldots k}$ those to the social network sample (SN) for the period (i. e., those who posted during the observation period). The resulting correction factor for representatives of that subpopulation is shown in expression (1).

$$\omega_{ij\ldots k} = (\%P^{R}_{ij\ldots k})/(\%P^{SN}_{ij\ldots k}) \tag{1}$$

A value less than 1 in expression (1) indicates that the subpopulation is overrepresented in the social network sample; otherwise, group underrepresentation is present. From this moment on, users are weighted according to the inverse of the previous ratio when included in spatial and/or temporal aggregations, considering their recent publications. Re-weighting as indicated would get us closer to the Mexican's mood project. However, the required information is not generally available to us. Another unresolved issue concerns the uncertainty introduced when "experts" tag the messages. In general, one message does not receive the same label when looked at by different individuals. We are not aware that some way to account for this uncertainty in algorithm training is being studied. However, asking the author of the publication directly about his state of mind at the time of writing could help reduce uncertainty. This does not fully eliminate uncertainty, since the recall process itself lacks certainties. However, there are other ways to reach out to Twitter users that reduce some risks. We will comment on them later as part of our proposal.

1.3 Prediction of Required Information

As indicated, our database of georeferenced tweets does not include information on age, sex, academic level, or socioeconomic status. Therefore, we are prevented from re-weighting tweets, as has been suggested. On Twitter, users can decide to keep their profiles confidential. Consequently, we are unable to access such information.

The second obvious option is to estimate the characteristics of interest of Twitter users. Answers by authors to a survey questionnaire can be used as tags, rather than resorting to experts. Assume that, over the interview, identify respondents in the "small sample", who are also recent network users. We seek to link each user's public messages with his/her sociodemographic information recorded in the questionnaire. It has been pointed out that informant consent to proceed with this linkage should be requested.[2] Respondents could provide their usernames or post a message containing a personalized code known only to us[3] as consent expression. Therefore, after a brief explanation of our objective, they will be made aware that if consent is granted, their recent messages will be labeled with their SD characteristics and treated with the same protections as the rest of the information they provided. Respondents are free to heed our request. If and when they do, usernames will be made available to us. We can now access their recent public posts and label them using socioeconomic information from the survey. In this fashion, training and testing sets will be created. The former will become input for training AI algorithms. Once the training results are evaluated through the testing set, algorithms deemed suitable may be used to automatically tag messages from recent Twitter users not in the sample. All tagged messages by the same user will be pooled to help decide how to tag the person since there is no guarantee that all messages by an author will get the same labels. In this fashion, we end up with two databases, one of tagged messages, and another of tagged users, which we shall refer to as the "large sample." The latter becomes the production database. In our proposal, the random sample is used to train algorithms, rather than to produce statistical results. If consent is not granted, we may face additional selection bias due to non-response. However, the survey itself provides sufficient information to compare what should have been, and what was finally received. If there are enough favorable responses from representatives of each defined class, re-weighting what was received is possible. It must be stressed that we have found no evidence that re-weighting is necessary or useful to train an algorithm, nor do we know of results in the literature that tell us how to proceed in the training phase.[4] In general, whether it is the labeling of the user database or the monitoring of a subject studied in a survey, the proposed roadmap to optimize the combination of survey and network data must include the following steps, as a minimum:

- First stage:
  - Select a random sample of a few thousand households or individuals. To be termed the *small sample*.
  - Collect topic and sociodemographic (SD) data.

[2] Although in a different context, see [1, 8], and [12].

[3] Which, by the way, would reduce typing errors.

[4] Though in a different context, see Jean Opsomer, IASS Webinar 29: Fitting Classification Trees to Complex Survey Data, 31 May 2023. http://isi-iass.org/home/wp-content/uploads/IASS-webinar-Opsomer.pdf.

- Identify social media users in the household and ask for their usernames.
- For each user, tag recent tweets with his/her survey information.
- Train and evaluate algorithms.
- Choose those that are appropriate.

– Second stage:

- Automatically tag all recent public posts at your disposal.
- Collect all tagged posts from the same user.
- Tag all users with recent posts. This is termed the *large sample*.
- Use their SD tags to reweight and allow for selection bias.
- Compute rates from SD corrected large sample.
- Publish daily, weekly, monthly, etc., results whenever precision allows, until next survey round

As an example, we can then use the tagged-users database to reduce selection bias in the positivity index in the "Mexican Twitter user's mood" as follows. Historic publications already tagged for sentiment will be used to identify users who posted on a certain period (day, week, or month). An author's sentiment is determined based on the positivity of all his/her posts for the period. Then, the relative size of each of the classes or subpopulations represented in the sample of users will be determined as a percentage of the total number of authors for each period. From expression (1), the weighting that will mitigate the selection bias in the results will be determined. The positivity ratio will be calculated as the ratio between the sum of weights of positive Twitter users, divided by the corresponding weighted sum of negative ones. A similar procedure will be followed to update results through daily capture of tweets.

1.3.1 Beyond SD Correction

The procedure described for the "Mexican Twitter users' mood" provides us with a way forward to produce statistical information with a smaller selection bias on other topics, based on the joint use of specialized surveys and social media posts. The lack of a sentiment survey is, however, an important difference. In many other instances, the magnitude of the selection bias can be gauged by comparing unbiased survey estimates with those obtained from Twitter posts, both weighted and unweighted.

2 Continuous Monitoring of Social Issues

Whenever linking is possible, respondents' tweets are tagged using survey responses. Tagged texts published on dates close to the interview will allow training of one or more algorithms covering topics dealt with by the survey. Of course, some will give better predictions than others. Automatic labeling through

Survey	Random sample's answers	Identify respondents who are $\mathbb{X}$ users	Tag their posts $P(S_j^{\mathbb{X}})$, form training (T) and testing (t) sets. Socio-demographic (SD) variables, too	Train and test algorithms. Use sampling design?	Tag all posts in time interval, including SD	Tag users, to form reweighted database	Infer to whole population by topic or across them	
Topic 1	$Q(S_1)$	$S_1^{\mathbb{X}} = S_1 \cap \mathbb{X}$	$T_1(S_1^{\mathbb{X}}) = Q(S_1^{\mathbb{X}}) \otimes P(S_1^{\mathbb{X}})$ $T_1 = T_{1T} \cup T_{1t}$	$Q_1(S_1^{\mathbb{X}}) = \mathcal{A}_1(P(S_1^{\mathbb{X}}))$	$\hat{Q}_1(P(\mathbb{X})) = \mathcal{A}_1(P(\mathbb{X}))$	$\hat{Q}(\mathbb{X}^{(\omega)}) = (\mathcal{W}_{SD}, \hat{Q}_1(\mathbb{X}), \hat{Q}_2(\mathbb{X}), \ldots \hat{Q}_k(\mathbb{X}))$	$\hat{Q}_1(\mathbb{X}^{(\omega)})$	$\hat{Q}(\mathbb{X}^{(\omega)})$
Topic 2	$Q(S_2)$	$S_2^{\mathbb{X}} = S_2 \cap \mathbb{X}$	$T_2(S_2^{\mathbb{X}}) = Q(S_2^{\mathbb{X}}) \otimes P(S_2^{\mathbb{X}})$ $T_2 = T_{2T} \cup T_{2t}$	$Q_2(S_2^{\mathbb{X}}) = \mathcal{A}_2(P(S_2^{\mathbb{X}}))$	$\hat{Q}_2(P(\mathbb{X})) = \mathcal{A}_2(P(\mathbb{X}))$		$\hat{Q}_2(\mathbb{X}^{(\omega)})$	
...	...	...	...	...	...		...	
Topic k	$Q(S_k)$	$S_k^{\mathbb{X}} = S_k \cap \mathbb{X}$	$T_k(S_k^{\mathbb{X}}) = Q(S_k^{\mathbb{X}}) \otimes P(S_k^{\mathbb{X}})$ $T_k = T_{kT} \cup T_{kt}$	$Q_k(S_k^{\mathbb{X}}) = \mathcal{A}_k(P(S_k^{\mathbb{X}}))$	$\hat{Q}_k(P(\mathbb{X})) = \mathcal{A}_k(P(\mathbb{X}))$		$\hat{Q}_k(\mathbb{X}^{(\omega)})$	

Fig. 1 Monitoring social issues in Mexico, linking social networks and surveys

algorithms trained using different surveys will result in multiple additional tags for each tweet in our historical database and those captured later, as well as for their authors. In this way, a survey such as the National Survey of Victimization and Perception of Public Safety (ENVIPE) will provide labels on crime and its causes or impact on victims. Based on their publications, these tags will be automatically assigned to authors who publish on dates close to and after the survey, for frequent statistical monitoring of these topics. Similarly, labor force surveys will follow the evolution of occupation. For the Mexican case, the use of rotating panels opens the possibility of studying how a person whose occupational status changes while in the sample tweets. Many topics, such as violence and crime, or emotional, economic, patrimonial, physical, and violence against women, can also be given constant attention. Under these conditions, and after re-weighting, it will be possible to carry out frequent and representative statistical monitoring of the topics studied by each survey. Figure 1 summarizes the aforementioned process. Each row refers to a survey on a specific topic. Information collected by the $i-th$ survey from sample S_i is stored in a database denoted by $Q(S_i)$. We ask each respondent to identify household members who are SN users. We do so once the interview ends, so our request will not affect respondent behavior. Next, we form the new set $S_i^{\mathbb{X}} = S_i \cap \mathbb{X}$ of survey respondents who are also SN users and agreed to share usernames with us. Note that two information sources are available for each of these individuals. Their answers to the survey, $Q(S_i^{\mathbb{X}})$, and their SN posts, which we denote by $P(S_i^{\mathbb{X}})$. All user's posts are tagged using the same answers to the survey; the result is denoted by $T_i(S_{iT}^{\mathbb{X}}) = Q(S_i^{\mathbb{X}}) \otimes P(S_i^{\mathbb{X}})$. T_i is split into training and testing sets T_{iT} and T_{it}, respectively. The first will be used to train algorithm A_i as in $Q_i(S_{iT}^{\mathbb{X}}) = A_i(P(S_{iT}^{\mathbb{X}}))$. The latter is used to evaluate the combination algorithm-tag-posts. From this point onwards, we no longer use the survey sample.

Suitable combinations are used to predict tags (i.e., answers to questionnaire items), for a wider population of SN users who posted during a given period, $\mathbb{X}$ say, as follows $\hat{Q}_i(P(\mathbb{X})) = A_i(P(\mathbb{X}))$. This yields a new database $T_i(P(\mathbb{X}))$, which this time includes all of $\mathbb{X}$ posts, rather than merely those of $S_i^{\mathbb{X}}$. In contrast with $T_i(S_{iT}^{\mathbb{X}})$, where all posts by the same user get the same tags, those in $P(\mathbb{X}))$ may be tagged differently by an algorithm. Under these conditions, a rule must be introduced to tag users from their tagged posts, e.g., majority rule. This is particularly important for sociodemographic variables since they will be used to correct for selection bias. Once the SD weights ω and the tags for each user, $\hat{Q}_i(\mathbb{X})$, are determined, a database of individual responses, $\hat{Q}(\mathbb{X}^{(\omega)})$ say, we are in a position to begin producing statistical information for the population at large, from that of $\mathbb{X}$, which we nw call the *large sample*.

$$\hat{Q}(\mathbb{X}^{(\omega)}) = (\mathbb{W}_{SD}, \hat{Q}_1(\mathbb{X}), \hat{Q}_2(\mathbb{X}), \ldots, , \hat{Q}_k(\mathbb{X}))$$

We are therefore faced with the possibility of crossing variables between surveys, since each user will be classified according to different themes. This is equivalent to having them simultaneously answer all survey questionnaires, which is clearly unthinkable with current methodologies. In this fashion, topics from various surveys can now be related. The closest equivalent to this in official statistics would be collecting information on many topics from a few million respondents. The main difference is that we are not directly asking them, but predicting their answers using ML. For example, it is possible to think about monitoring the mental health of people who have lost their jobs or some other type of loss. Another example could be a change in consumer confidence for recent crime victims. In short, a wide range of possibilities opens.

Furthermore, various surveys allow identification of changes in the immediate environment of the user-respondent, such as the birth of a new household member, the return of a migrant, or the marriage of a relative, which can influence how the user expresses himself on the net. We would like to think that the use of all or part of this information as labels associated with messages will allow new algorithms to be trained with reasonable precision. This would lead us to relate posts to events not directly associated with the user-respondent.

3 Final Comments

Development of the present proposal began some years ago, as can be surmised from the use of the term Twitter, rather than that net's new designation: X. Overtime, network policy regarding free access to a fraction of public posts was canceled. INEGI, for one, stopped collecting such posts. Therefore, we have been unable to test our proposal in the field and cannot show any evidence of its viability. Hence, the use of the term "Blueprint" in the title of this chapter. We remain optimistic, however, that its introduction in the official statistics production process may save time and money in the production of sociodemographic statistical information. The accumulated experience in the use of information from social networks has been invaluable for its exploitation in the production of official statistics. The group of technicians and professionals at INEGI who have been trained in the use of the relevant techniques has already reached a significant number. Likewise, the plans to strengthen the infrastructure show significant progress. In our proposal, we have outlined a way forward for continuous monitoring of social and demographic issues of interest to national statistical offices. This is supported by the information collected within the traditional programs to produce official statistical information through sample surveys. In this way, the selection bias is reduced in the indicators with which various topics will be monitored continuously. Likewise, the uncertainty behind these indicators is reduced by avoiding relying on expert opinion for the labeling of training and testing sets. Reiterating, whether it is the labeling of the user database, or the monitoring of a subject studied in a survey, the proposed roadmap to optimize the combination of survey and network data must include the steps summarized in Sect. 1.3. Adapting our proposal to input from administrative records for similar purposes is not immediate. For this case, ways will have to be developed to link information in administrative records with the citizen's publications. However, it will be necessary to be cautious to avoid making public information whose inappropriate use by third parties could harm our respondents. For example, we suggest avoiding the use of the Unique Population Registry Code (CURP) used in Mexico for official purposes, but whose publication in a tweet would allow linking the information from various public registries with posts from the social network.

References

1. Baghal, T., Sloan, L., Jessop, C., Williams, M. L., and Burnap, P.: Linking Twitter and Survey Data: The Impact of Survey Mode and Demographics on Consent Rates Across Three UK Studies, Social Science Computer Review, Vol. 38(5) 517–532, (2020).
2. Bustos, V. A., Coronado, A. A., Fraustro, S. L., Leyva, G., López, N., Olvera, R. A., Romo, A. M., Silva, V.: Caracterización del sesgo de selección en redes sociales en México a través de algunas características sociodemográficas de sus usuarios, Realidad, Datos y Espacio, Revista Internacional de Estadística, Vol. 13, Núm. 2, 26–43, (2022). https://rde.inegi.org.mx/index.php/2022/07/06/caracterizacion-del-sesgo-de-seleccion-en-redes-sociales-en-mexico-a-traves-de-algunas-caracteristicas-sociodemograficas-de-sus-usuarios/
3. Data-Pop Alliance: Opportunities and Requirements for Leveraging Big Data for Official Statistics and the Sustainable Development Goals in Latin America. Data-Pop Alliance (Harvard Humanitarian Initiative, MIT Media Lab and Overseas Development Institute). (2016).
4. Destatis: Access to Big Data for statistical purposes, Note by the Federal Statistical Office of Germany, Economic Commission for Europe, Conference of European Statisticians, 67th plenary session, Paris, 26–28 June, (2019). https://undocs.org/ECE/CES/2019/20
5. Golub, B. and Jackson, M. O.: Using selection bias to explain the observed structure of Internet diffusions, Proceedings of the National Academies of Sciences, June 15, vol. 107, no. 24, 10833–10836, (2010) www.pnas.org/cgi/doi/10.1073/pnas.1000814107

6. Iacus, S., Porro, G., Salini, S., Siletti, E.: Controlling for Selection Bias in Social Media Indicators through Official Statistics: a Proposal, Journal of Official Statistics, Vol. 36, No. 2, (2020), pp. 315–338, http://dx.doi.org/10.2478/JOS-2020-0017
7. Jansen, R.: UN Global Working Group (GWG) on Big Data and its Task Teams, International Symposium on the Use of Big Data for Official Statistics, National Bureau of Statistics of China, Hangzhou, China, Oct. 16–18, (2020). http://www.stats.gov.cn/english/pdf/202010/P020201012399997943871.pdf
8. Kreuter, F., Sakshaug, J. W., and Tourangeau, R.: The Framing of the Record Linkage Consent Question, International Journal of Public Opinion Research, Vol. 28, No. 1, (2016).
9. Lokanathan, S., Perera-Gomez, T., Zuhyle, S.: Mapping Big Data Solutions for the Sustainable Development Goals [Draft], LIRNEasia, (2017). https://lirneasia.net/2017/03/mapping-big-data-solutions-sustainable-development-goals/
10. Smith, H.: Big Data for Official Statistics, Workshop on Big Data for Economic Statistics: Challenges and Opportunities, 11 September 2018, Rio de Janeiro, Brazil, https://unstats.un.org/Unsd/nationalaccount/workshops/2018/rio/UNSD.PDF
11. Snyder, N., UN Global Working Group on Big Data, UNECE Workshop on Statistical Data Collection, Washington, DC, 29 April – 1 May, (2015)
12. Stier, S., Breuer, J., Siegers, P., and Thorson, K., Integrating Survey Data and Digital Trace Data: Key Issues in Developing an Emerging Field, Social Science Computer Review, Vol. 38(5) 503–516, (2020).
13. Struijs, P., Braaksma, B., and Daas, P., Official statistics and Big Data, Big Data & Society, April–June 2014: 1–6, The Author(s) (2014), doi: 10.1177/2053951714538417, https://journals.sagepub.com/doi/abs/10.1177/2053951714538417
14. Struijs, P., Daas, P., Quality approaches to big data in official statistics, European Conference on Quality in Official Statistics, (2014). http://www.pietdaas.nl/beta/pubs/pubs/Q2014_session_33_paper.pdf
15. Stark, T. H.: Understanding the selection bias: Social network processes and the effect of prejudice on the avoidance of outgroup friends. Social Psychology Quarterly, 78(2), 127–150, (2015). https://doi.org/10.1177/0190272514565252
16. Tam, S. M. and Kim, J. K.: Big Data ethics and selection-bias: An official statistician's perspective, Statistical Journal of the IAOS 34, 577–588, IOS Press, (2018). DOI: 10.3233/SJI-170395
17. Van Halderen, G., Bernal, I., Sejersen, T., Jansen, R., Ploug, N., Truszczynski, M., Big Data for the SDGs, Country examples in compiling SDG indicators using non-traditional data sources, Working Paper Series, SD/WP/12/January 2021, ESCAP Statistics Division (2021). https://www.unescap.org/sites/default/d8files/knowledge-products/SD_Working_Paper_no12_Jan2021_Big_data_for_SDG_indicators.pdf
18. UNSD, Report of the Global Working Group on Big Data for Official Statistics, Statistical Commission Forty-sixth session, 3–6 March, 2015, New York (2015). https://documentSDds-ny.un.org/doc/UNDOC/GEN/N14/692/71/PDF/N1469271.pdf?OpenElement

Dynamic Evaluation of Electoral Preferences in Mexico: A Bayesian Power Ranking Model

Michelle Anzarut, Luis Felipe González, and María Teresa Ortiz

Abstract We present a Bayesian Power Ranking model to analyze Mexico's electoral preferences. This model ranks potential presidential candidates for the 2024 elections using a wide array of surveys, each including a different subset of the roster of plausible candidates. The model employs a multinomial approach for observations, corrects for polling house biases, and uses a correlated random walk to dynamically update preferences. We defined the Potential Index, an index of relative preference probabilities, and established the Power Ranking, a comprehensive preference order over time. Throughout 2023, both the Potential Index and the Power Ranking provided insights into public opinion and media influence before the announcement of the official candidate list.

Keywords Bayesian inference · Dynamic models · Electoral preferences · Preelection polls · Ranking system

1 Introduction

This chapter describes a Bayesian dynamic model used to rank the preferences over possible candidates for the 2024 presidential elections in Mexico. From mid-2021 onward, various parties unofficially announced their prospective candidates. The evolving political landscape made the list of candidates highly fluid; names were regularly added or removed based on political strategies and public sentiment. Several political figures declared their intentions to run, while polling houses identified others as likely candidates based on their perceived competitiveness.

M. Anzarut (✉)
Universidad Nacional Autónoma de México, Mexico City, Mexico
e-mail: michelle@sigma.iimas.unam.mx; anzarutm@hotmail.com

L. F. González · M. T. Ortiz
Instituto Tecnológico Autónomo de México, Mexico City, Mexico
e-mail: luis.gonzalez@itam.mx; maria.ortiz@itam.mx

J. A. Christen et al. (eds.), *Statistics, Society and Environment*, Springer Proceedings in Mathematics & Statistics 479, https://doi.org/10.1007/978-3-031-78401-9_4

Polling agencies conducted surveys comparing two or more potential presidential candidates with questions such as, "Among the following people, who do you prefer as the candidate for the 2024 presidential elections in Mexico?" and "If today were the election day, who would you vote for?" We defined potential candidates as those who had been featured in these types of surveys since 2021. Our ranking, regularly updated, synthesized the outcomes of numerous such polls until the announcement of the official candidate list.

Our proposal employs a multinomial approach to calculate voter preferences, adjusting for polling house biases. Then, it updates the preference states dynamically to accurately reflect changes in the precandidates popularity, facilitating a thorough analysis of electoral dynamics. The model gave an ordering of electoral preferences at distinct moments in time. Hence, we named it the Power Ranking model (PR model). Furthermore, it also provided an index that allowed the differences in preferences to be measured on a numerical scale. When candidates and parties are already established, the normalized index can serve as a Poll of Polls model.

The structure of this chapter is organized as follows. Section 2 introduces the PR model, discussing its motivation and definition. Section 3 focuses on the data used and challenges related to data comparability. The methodology for estimating parameters and optimizing the model's performance is elaborated in Sect. 4. Section 5 shows the practical application of the model. The chapter concludes in Sect. 6 with final remarks and potential directions for future research.

2 Model

This section delves into the Power Ranking (PR) model. Section 2.1 explains the background that led to its definition. We then define the model, which is bifurcated into two main parts. Section 2.2, the observation model, focuses on processing preference estimates at a specific point in time for each potential candidate. Section 2.3, the preference model, then models the dynamic nature of the candidates' preferences between themselves and over time, using a correlated random walk. Finally, Sect. 2.4 explains the index and ranking determination.

2.1 Background

Our ranking proposal is a Bayesian and dynamic version of the Bradley–Terry model [2]. The Bradley–Terry model is a probabilistic framework for analyzing paired comparisons. Each item in a pair is assigned a positive strength parameter. For two items A and B with strength parameters p_A and p_B respectively, the model defines the probability of A winning over B as

$$P_{A,B} = \left(\frac{p_A}{p_A + p_B}\right).$$

This results in a model that generates independent binomial realizations between pairs of items. The model finds extensive application in various fields for ranking and preference analysis, owing to its straightforward methodology in quantifying and comparing the relative strengths of two items. The Bradley–Terry model has been extended in several ways. Leonard [5] introduced a Bayesian approach, reparametrizing with $\theta_A = \log(p_A)$ and $\theta_B = \log(p_B)$, and employing multivariate normal distributions for the transformed parameters. Therefore,

$$P_{A,B} = \left(\frac{\exp(\theta_A)}{\exp(\theta_A) + \exp(\theta_B)}\right).$$

The latter leads to a more flexible approach, readily allowing for prior relationships between the parameters and permitting meaningful hierarchical prior distributions.

Such Bayesian extension can also be used with dynamic parameters,

$$P_{A,B}(t) = \left(\frac{\exp(\theta_A(t))}{\exp(\theta_A(t)) + \exp(\theta_B(t))}\right).$$

For example, in [3], the team ratings, $\theta_A(t)$ and $\theta_B(t)$, progress over time following an AR(1) model.

Suppose we now have m teams and wish to extend the previous model to define a set of estimated competitor skills for each time t, denoted as $(q_1(t), \ldots, q_m(t))$. We can use the softmax function for this extension,

$$(q_1(t), \ldots, q_m(t)) = \mathsf{softmax}\left((\theta_1(t), \ldots, \theta_m(t))\right),$$

where

$$\mathsf{softmax}((z_1, \ldots, z_p))_j = \frac{\exp(z_j)}{\sum_{i=1}^{p} \exp(z_i)}.$$

And the probability of a specific set of results in n competitions has a multinomial distribution, that is,

$$(x_1(t), \ldots, x_m(t)) \sim \textbf{Multinom}\left((q_1(t), \ldots, q_m(t)), n\right).$$

Here, $x_i(t)$ denotes the number of times the competitor i is declared the winner in the n events.

Let us translate this model to the possible candidates preference setting. Suppose we observe the results of a survey at one point in time t. Let m be the number of precandidates in the survey, let n be the sample size of the survey, and let $n_i(t)$

denote the number of times the competitor i wins in the survey. Then we can assume that:

$$(n_1(t), \ldots, n_m(t)) \sim \mathbf{Multinom}\left((q_1(t), \ldots, q_m(t)), n\right),$$

with

$$(q_1(t), \ldots, q_m(t)) = \mathsf{softmax}\left((\theta_1(t), \ldots, \theta_m(t)\right).$$

There are three factors that have not yet been taken into account in the previous model. First, we want to have a correction term for the sample size that allows us to make adjustments for the differences between the declared sampling error and the actual error. A historical analysis of electoral surveys conducted from 2018 to 2021 reveals that the actual errors in the polls are, on average, two times greater than the reported errors. This significant discrepancy indicates that the declared sampling errors underrepresent the true variability in the polling data. To address this issue, we introduce a correction factor $\phi \in (0, 1]$, which serves to discount the original sample size, thereby recalibrating the influence of each sample. Consequently, the adjusted sample size is calculated as $n' = \phi n$. A minor adjustment needs to be made since $n'_i = \phi n_i$ may result in a fraction for some i, in that case we take n'_i as the floor of the number.

The second factor is that we notice that the outcome probability depends solely on the latent strengths of the candidates being compared. In practice, other factors might affect the outcome. Krese and Štrumbelj [4] account for these covariates with a linear term. Following a similar idea, we will add a term of biases. This term allows us to consider advantages of polling houses for certain candidates and estimate these advantages from the data.

The third factor is that surveys do not face all the possible candidates. In fact, each survey includes a different subset of the roster of plausible candidates. Let us denote with $(d_1, \ldots, d_m)$ to the precandidates in this survey, let M be the total number of precandidates, and let $(c_1, \ldots, c_M)$ be the whole set of precandidates, then, there exists a function $f : \{1, \ldots, m\} \mapsto \{1, \ldots, M\}$ such that

$$d_k = c_{f(k)}, \quad \text{for } k = 1, \ldots, m.$$

Considering this function, we can write the model as,

$$\left(n'_{f(1)}(t), \ldots, n'_{f(m)}(t)\right) \sim \mathbf{Multinom}\left(\left(q_{f(1)}(t), \ldots, q_{f(m)}(t)\right), n'\right),$$

with

$$\left(q_{f(1)}(t), \ldots, q_{f(m)}(t)\right) = \mathsf{softmax}\left(\left(\theta_{f(1)}(t), \ldots, \theta_{f(m)}\right) + \left(\alpha^{\text{house}}_{f(1)}, \ldots, \alpha^{\text{house}}_{f(m)}\right)\right).$$

Here α_k^{house} is a random variable that denotes the bias of the polling house who did the survey for the precandidate k.

Note that, $(\theta_1(t), \ldots, \theta_M(t))$ is the vector of preferences at time t in the logistic scale. This preferences will be modeled in time using a correlated random walk. The formal definition of the model is presented in the next section.

2.2 *Observations Model*

Let M be the number of potential presidential candidates for the Mexican elections of 2024, and let $(c_1, \ldots, c_M)$ denote these precandidates. Suppose we observe the results of a survey at a specific point in time t. Let $m \leq M$ represent the number of candidates included in the survey, and let $(d_1, \ldots, d_m)$ be these surveyed precandidates. Denote the survey's sample size as n. We define a mapping function $f : \{1, \ldots, m\} \mapsto \{1, \ldots, M\}$ such that

$$d_k = c_{f(k)}, \quad \text{for } k = 1, \ldots, m.$$

We denote the estimated preferences in the survey as $(p_{f(1)}, \ldots, p_{f(m)})$ and assume that

$$\left(\lfloor \phi n p_{f(1)}(t) \rfloor, \ldots, \lfloor \phi n p_{f(m)}(t) \rfloor\right) \sim \textbf{Multinom}\left(q_f(t), n'\right), \tag{1}$$

where $n' = \sum_{i=1}^{m} \lfloor \phi n p_{f(i)}(t) \rfloor$ is the adjusted sample size, discounted by a fixed value, and $q_f(t) := (q_{f(1)}(t), \ldots, q_{f(m)}(t))$ is characterized as

$$q_f(t) = \mathsf{softmax}\left(\theta_f(t) + \alpha_f^{\text{house}}\right). \tag{2}$$

Here, $\theta_f(t) := \left(\theta_{f(1)}(t), \ldots, \theta_{f(m)}\right)$, and $\alpha_f^{\text{house}} := \left(\alpha_{f(1)}^{\text{house}}, \ldots, \alpha_{f(m)}^{\text{house}}\right)$, where α_k^{house} denotes the bias of the polling house that conducted the survey for the precandidate k.

We assume that $\alpha_1^{\text{house}}, \ldots, \alpha_m^{\text{house}}$ are independent identically distributed normal random variables with a fixed standard deviation $\sigma_{\alpha^{\text{house}}} > 0$. This assumption implies that while biases are potentially impactful, they are centered around zero. This assumption is grounded in empirical observations from a historical analysis of polling data spanning the 2018–2021 electoral polls in Mexico.

For simplicity, we are omitting the dependence of the parameters on the survey. However, it should be noted that m, n, n', f, the preferences, and α^{house} are dependent on the survey, while the discount factor ϕ is the same for all surveys.

Finally, $\theta(t) := (\theta_1(t), \ldots, \theta_M(t))$ is the vector of preferences at time t on the logistic scale, adjusted for the effects of the polling houses. This vector is our primary parameter of interest, as it will define the ranking.

2.3 Preference Model

We share information between the precandidates at the same time and also over time. We achieve this by modeling the preferences as a correlated random walk:

$$\theta_k(t) = \theta_k(t-1) + \lambda_k(t), \; k = 1, \ldots, M, \tag{3}$$

where $\theta_k(1) = \theta_k^{\text{init}}$ for $k = 1, \ldots, M$, are independent identically distributed standard normal random variables, and $\lambda_k(t)$ is a change at time t in the level of the precandidate's preference. These level changes are modeled with a multivariate normal distribution.

$$\lambda(t) := (\lambda_1(t), \ldots, \lambda_M(t)) \sim \mathcal{N}_M(0, \Sigma). \tag{4}$$

Decomposing $\Sigma = D_\psi \Omega D_\psi$, where D_ψ is the diagonal matrix with positive entries $(\psi_1, \ldots, \psi_M)$ and Ω is the correlation matrix, we assume that $\psi_k \sim \mathcal{N}(0, \sigma)$ for $k = 1, \ldots, M$ and $\Omega \sim \text{LKJCorr}(2) \propto |\Omega|$ (See [1] and [6]). The fixed value σ controls how dynamic the model.

Finally, let $\bar{\theta}(t)$ denote the mean of $\theta(t)$. For each time t, the Potential Index and the Power Ranking are determined by the posterior distribution of the centered preferences $\theta^*(t) := \theta(t) - \bar{\theta}(t)$. We subtract the mean to enhance the identifiability of the model. The details are explained in the next subsection.

2.4 Potential Index and Ranking Determination

The Potential Index is defined as the 10th percentile of the posterior distribution of centered preferences, $\theta^*(t)$. This metric differs from traditional methods that use the average or expected value of the posterior distributions, offering a more conservative assessment of a candidate's standing.

The 10th percentile usage aligns with an asymmetric loss function,

$$L(x, \hat{x}) = \begin{cases} a(\hat{x} - x) & \text{if } \hat{x} > x, \\ b(x - \hat{x}) & \text{if } \hat{x} \leq x, \end{cases}$$

where $a > b$, indicating a higher penalty for overestimations. This measure is taken to mitigate the risk of overestimating a candidate's preference, particularly in short-term media-driven scenarios. Focusing on the lower end of the preference spectrum, the Potential Index reduces the influence of outliers and brief surges in popularity.

When pollsters begin to overlook a candidate, leading to increased uncertainty, their index naturally decreases. Thus, the Potential Index reflects the attention pollsters pay to candidates, balancing the dynamics of public opinion and media influence in their assessment.

The Potential Index is multiplicatively interpreted, indicating a candidate's relative preference probability at a given time. For instance, if in July, candidate A has an index of 300 and B has 150, it suggests that A is twice as likely to be preferred over B. Absolute differences in these numbers are not insightful regarding preference levels. Moreover, a change in a candidate's index over time, such as A's increase to 200 in August, doesn't imply a rise in popularity, as indices are comparable only among candidates within the same period.

Finally, we define Power Ranking as the ordering of candidates based on the Potential Index.

3 Data

The observed data are surveys, each detailing the polling house, research methodology, publication date, sample size, reported error, and the number of precandidates (ranging from 2 to 8). Preferences obtained were also recorded, excluding responses such as "Does not know," "Did not answer," or "None." The last published results of the model were computed with more than 3000 surveys from 40 polling houses, spanning from September 2021 to September 2023. The margin of sampling errors ranges from 0.5 to 5% points (surveys with a tiny reported error are usually performed online).

A major obstacle is the comparability of surveys. Despite having variables to examine, such as research methodology, sample size, or reported error, their interrelationship is unclear. We considered categorizing surveys into tiers based on methodology (e.g., telephone vs. online surveys) and adjusting sample sizes accordingly. However, this method was not used due to two factors: the absence of real calibration data and the inconsistency in reported versus actual sampling errors found in other Mexican voting polls. Analyzing other voting exercises we find an average error of about 5% points, regardless of smaller reported errors.

The role of sampling error in rankings is another crucial consideration. A large gap between candidates minimizes the impact of sampling error, while a small gap makes the ordering highly sensitive to these errors. Addressing these complexities, we adopted a conservative approach by considering a minimum error of 7% points for estimates around 50%, corresponding to an effective sample size of approximately 200 interviews. In scenarios where the gap between candidates is small, the reduced effective sample size results in a less separation between precandidates.

4 Model Estimation and Performance Optimization

The model is written in Stan code [9], and the data fitting is straightforward using the CmdStanR package from R [8]. See, for example, [10].

We adhered to the Bayesian workflow outlined in [7]. We began by ensuring that the chosen priors were both sound and aligned with established domain knowledge. We then utilized the diagnostic tools provided by Stan to validate our model's fitting procedure. Subsequently, we assessed the model's adequacy, confirming its accurate representation of the observed data. In Sect. 4.1, we delve into this last step, focusing on parameter calibration and model evaluation using part of the data for training, and later surveys for predictions. Section 4.2 introduces a model optimization. This optimization was designed to mitigate the impact of transient media attention, distinguishing it from the consistent support that other potential candidates receive in the ranking.

4.1 Evaluation and Parameter Calibration

For parameter calibration and to evaluate the model's predictive capability, we divided the data into two sets. The training data consists of surveys from November 2021 to June 2022, while the prediction set includes surveys conducted post-June 2022.

The metrics we use for the model evaluation are ranking error and preference estimation error. Ranking error is the percentage of incorrect predictions when facing two candidates in a poll. The preference estimation error is the mean of the absolute differences between the preference percentages obtained when confronting two candidates in a survey and the preference estimated by the model. In case there are more than two candidates, we compute the possible pairs and normalize to effective preference. The Ranking error serves to evaluate the Power Ranking, and the preference estimation error is employed to assess the Power Index.

There are two parameters that we need to calibrate. The first one is ϕ, which discounts the sample size, so we call it the discount factor. The second one is σ, which controls how dynamic the model is. We call σ the dynamic factor. The higher σ is, the more preferences can move over time.

We ran the model for different values of both parameters (generating all possible combinations). Figure 1 shows the results. Consistently, the model performs better when the dynamic factor is high and the discount factor is low.

We opted for a discount factor of 0.125 and a dynamic factor of 0.15. This combination resulted in a Ranking error of 11.6% and a preference estimation error of less than 10%. We consider these error levels acceptable, rendering both the Power Ranking and the Power Index reliable for use.

4.2 Penalty Factor

The power ranking, derived from the PR model, captures two key aspects: media attention and the positions of possible candidates in the polls. Upon implementing

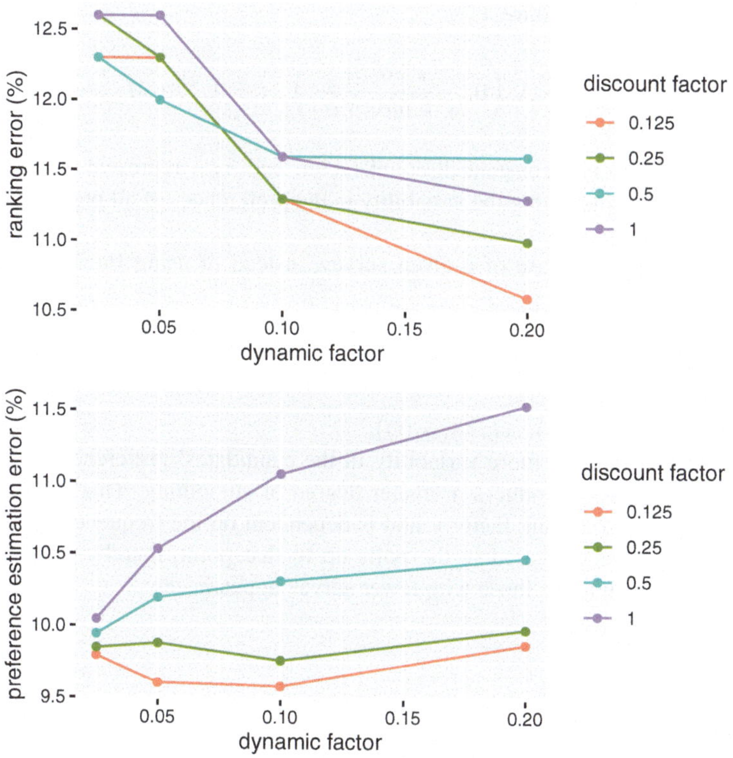

Fig. 1 The top panel shows the ranking error for different combinations of dynamic factors and discount factors. Ranking error is the percentage of incorrect predictions when facing two candidates in a poll. The bottom panel shows the preference estimation error for the same combinations of dynamic factors and discount factors. The preference estimation error is the mean of the absolute differences between the preference percentages obtained when confronting two candidates in a survey and the preference estimated by the model

the model, we observed that potential candidates can suddenly emerge in surveys and rapidly ascend in the rankings, often due to media events that briefly focus the spotlight on them. Our goal is to distinguish these temporary surges in visibility from the sustained, long-term support that more established candidates receive in the ranking. To do this, our analysis is limited to candidates who have appeared in at least two polls. Moreover, we have introduced a penalty factor to adequately address the uncertainty associated with candidates who have limited appearances in survey data.

Initially, we estimate the probability distribution of the vector of preferences at time t, $\theta(t)$, as outlined in Sects. 2.2 and 2.3. Then, we simulate a disturbance term, $\eta(t) := (\eta_1(t), \ldots, \eta_M(t))$. At each time point t and for every precandidate i, $\eta_i(t)$ is a random error from a normal distribution with mean 0, and its standard deviation is adjusted according to the survey frequency of precandidate i at that time, denoted by surv-$n_i(t)$. Thus, we have

$$\eta_i(t) \sim \mathcal{N}\left(0, \frac{\kappa \cdot \sigma_{\theta(t)}}{1 + \text{surv-n}_i(t)}\right), \; i = 1, \ldots, M,$$

where $\kappa \in (0, 1)$ is a constant discount factor, and $\sigma_{\theta(t)}$ represents the standard deviation of $\theta(t)$, indicating the variability in the preferences of all precandidates at a that specific time.

In the simulation phase of a given survey, instead of using Eq. 2, we add the disturbance term,

$$q_f(t) = \mathsf{softmax}(\theta_f(t) + \alpha_f^{\text{house}} + \eta_f(t)).$$

where $\eta_f(t) := \left(\eta_{f(1)}(t), \ldots, \eta_{f(M)}(t)\right)$.

Note that, if there is more variability in the candidates' preferences at a given moment, the model introduces a greater degree of smoothing. This smoothing is modulated by the discount factor κ and is dependent on the frequency with which each candidate has appeared in the polls up to that point, thereby reflecting the increased uncertainty for those with fewer survey appearances.

5 Results

In this section, we demonstrate the application of the model. Figure 2 displays the Potential Index's progression from early 2022 to the announcement of presidential candidates, tracking the indices of 76 individuals. Table 1 shows detailed data as of September 10, 2023, including each individual's name, Power ranking, and Potential Index.

To facilitate visualization and comparison, we will focus on the five individuals who were at the top of the rankings in the last calculation of the index on September 10th, 2023. The top panel of Fig. 3 shows the Potential Index for these individuals, while the bottom panel displays their Power Ranking. Various analyzes can come from these measurements. We can see, for example, that Claudia Sheinbaum and Marcelo Ebrard continued to alternate in the first two places in the ranking throughout 2022, although as of 2023, it was Claudia Sheinbaum who was clearly positioned in first place. We also see how the potential indices are, at first similar, and then they separate, with Claudia Sheinbaum taking an important advantage. Another interesting character is Xóchitl Gálvez, who abruptly goes from being below twentieth place in the Power Ranking for a long period, to second place.

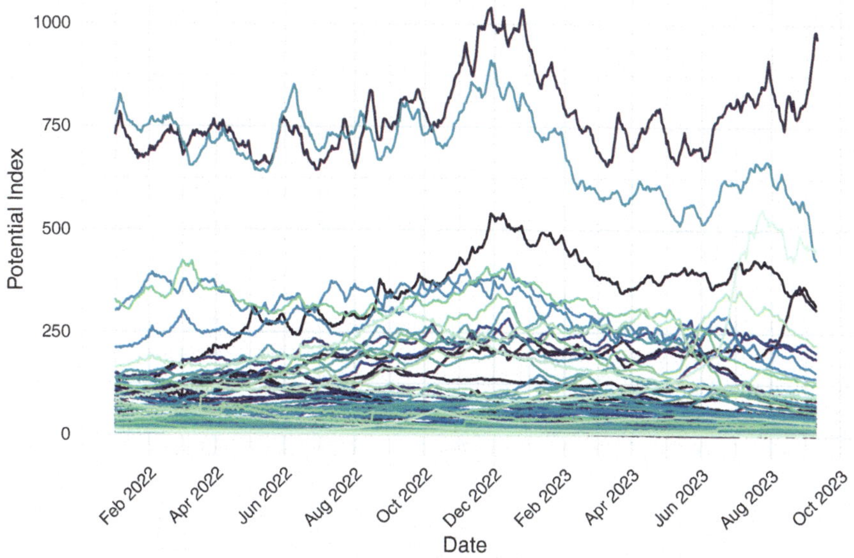

Fig. 2 Potential Index for 76 possible candidates from the beginning of 2022 until September 2023

6 Conclusions and Future Work

We introduced the PR model as a novel method for navigating the electoral landscape during periods where candidate identities remain undefined. This model utilized public surveys facing various individuals against each other. Some of these individuals were clear potential presidential candidates, while others were speculative choices by polling agencies. To the best of our knowledge, no tool in Mexico offered a concise way to track or compare the multitude of individuals featured in these numerous polls. The PR model addressed this need, its effectiveness was clear in two main metrics: the Potential Index, which offered a numerical reference for the relative preferences of each potential candidate, and the Power Ranking, which ordered these preferences.

The Power Ranking was well-received, often appearing in the headlines. However, the more informative Potential Index faced challenges because of its multiplicative nature, leading to less frequent use and occasional misinterpretations in the media. Unfortunately, while odds and other multiplicative measures are commonly used in sports, they are not as prevalent in Mexico's survey industry.

Future enhancements to the PR model might include adding predictive variables like media mentions or the social media activities of individuals. The aim would be to develop a model with covariates that can smooth out abrupt fluctuations caused by media coverage, thus reducing the need for perturbation adjustments during simulations.

Table 1 Power ranking (PR) and Potential Index (PI) for 76 possible candidates as of September 10, 2023

PR	Name	PI	PR	Name	PI
1	Claudia Sheinbaum	965	39	Silvano Aureoles	33
2	Xóchitl Gálvez	447	40	Ildefonso Guajardo	32
3	Marcelo Ebrard	427	41	Manuel Clouthier	32
4	Adán Augusto López	318	42	Diego Sinhue Rodríguez	30
5	Beatriz Paredes Rangel	307	43	Manuel Velasco	30
6	Santiago Creel	225	44	Alberto Anaya	28
7	Gerardo Fernández Noroña	203	45	Rocío Nahle García	26
8	Enrique de la Madrid	188	46	Armando Ríos Piter	25
9	Luis Donaldo Colosio Riojas	154	47	Beatriz Gutiérrez Müller	23
10	Lilly Téllez	141	48	Claudia Pavlovich	22
11	Samuel García	132	49	Jorge Emilio González	22
12	Ricardo Anaya	125	50	Salomón Chertorivski	22
13	Miguel Ángel Mancera	102	51	Ivonne Ortega	21
14	Tatiana Clouthier	98	52	Lázaro Cárdenas Batel	19
15	Rubén Moreira	93	53	Jorge Álvarez Máynez	17
16	Claudia Ruiz Massieu	89	54	Miguel Riquelme	17
17	Margarita Zavala	79	55	Patricia Armendáriz	17
18	Alfredo del Mazo	75	56	Esteban Moctezuma	16
19	Fco. García Cabeza de Vaca	73	57	Juan Ramón de la Fuente	16
20	Miguel Ángel Osorio Chong	69	58	Andrés Manuel López Beltrán	14
21	Alejandro Moreno	64	59	Javier Corral	13
22	Dante Delgado	60	60	Juan Zepeda	12
23	Mauricio Kuri	60	61	Claudio X. González	9
24	Ricardo Monreal	58	62	Hugo López-Gatell	9
25	Enrique Alfaro	51	63	Jesús Zambrano	9
26	José Ángel Gurría	50	64	Santiago Nieto	8
27	Manuel Velasco Coello	49	65	Ildelfonso Guajardo	7
28	Damián Zepeda	46	66	Ricardo Gallardo	7
29	Manlio Fabio Beltrones	45	67	César Verástegui	6
30	Alejandro Murat	44	68	Hugo López Gatell	5
31	María Eugenia Campos Galván	41	69	Quirino Ordaz	4
32	Omar Fayad	37	70	Rafael Marín	4
33	Juan Carlos Romero Hicks	36	71	Ulises Ruiz Ortiz	4
34	Patricia Mercado	36	72	Félix Salgado Macedonio	3
35	Cecilia Soto	35	73	Jaime Bonilla	3
36	Francisco Domínguez S.	34	74	Rocío Nahle	3
37	Gustavo de Hoyos	34	75	Yeidckol Polevnsky	3
38	Clemente Castañeda	33	76	Antonio Pérez Garibay	2

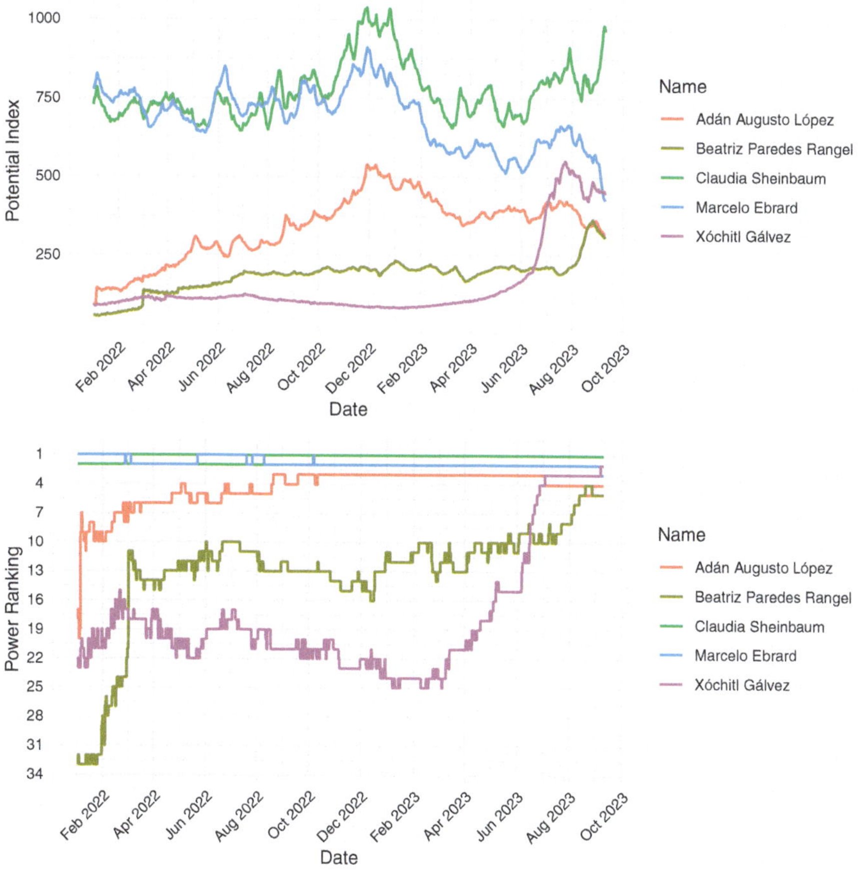

Fig. 3 The top panel shows the Potential Index for the five possible candidates who were at the top of the rankings in the last calculation of the index on September 10th, 2023, from the beginning of 2022 until September 2023. The bottom panel shows their Power Ranking

References

1. Barnard, J., McCulloch, R., Meng, X.L.: Modeling covariance matrices in terms of standard deviations and correlations, with application to shrinkage. Statistica Sinica pp. 1281–1311 (2000)
2. Bradley, R.A., Terry, M.E.: Rank analysis of incomplete block designs: I. The method of paired comparisons. Biometrika **39**(3/4), 324–345 (1952)
3. Glickman, M.E.: Paired comparison models with time-varying parameters. Harvard University (1993)
4. Krese, B., Štrumbelj, E.: A Bayesian approach to time-varying latent strengths in pairwise comparisons. Plos one **16**(5), e0251945 (2021)
5. Leonard, T.: An alternative Bayesian approach to the Bradley-Terry model for paired comparisons. Biometrics pp. 121–132 (1977)

6. Lewandowski, D., Kurowicka, D., Joe, H.: Generating random correlation matrices based on vines and extended onion method. Journal of multivariate analysis **100**(9), 1989–2001 (2009)
7. Schad, D., Betancourt, M., Vasishth, S.: Toward a principled Bayesian workflow in cognitive science. Psychological Methods **26** (2020). https://doi.org/10.1037/met0000275
8. Stan Development Team: CmdStanR: the R interface to CmdStan (2021). URL https://mc-stan.org/r-packages/
9. Stan Development Team: Stan Modeling Language Users Guide and Reference Manual, Version 2.27. Available from: https://mc-stan.org (2021)
10. Team, S.D.: Bradley-terry model for team comparisons. https://github.com/stan-dev/example-models/blob/master/knitr/bradley-terry/team.stan (2024)

Modeling Oral Glucose Tolerance Test (OGTT) Data and Its Bayesian Inverse Problem

Nicolás Kuschinski, J. Andrés Christen, Adriana Monroy, and Silvestre Alavez

Abstract One common way to test for diabetes is the Oral Glucose Tolerance Test or OGTT. Most common methods for the analysis of the data on this test are wasteful of much of the information contained therein. We propose to model blood glucose during an OGTT using a compartmental dynamic model with a system of Ordinary Differential Equations (ODEs). Our model works well in describing most scenarios that occur during an OGTT considering only four parameters. Fitting the model to data is an inverse problem, which is suitable for Bayesian inference. Priors are specified, and posterior inference results are shown using real data.

Keywords Bayesian analysis · Inverse problems · Regression · Statistics in medicine

1 Introduction

Type 2 diabetes is a serious and common illness. It typically develops without obvious symptoms and can go undiagnosed for years. With timely diagnosis and treatment, type 2 diabetes can change from a serious and potentially life-threatening illness to a relatively mild condition. Similarly, a patient who is informed that he or she is at risk of developing type 2 diabetes can often take appropriate measures and thus prevent diabetes from occurring altogether [11, 14].

Diabetes occurs when the body is unable to adequately regulate blood glucose. The body's main mechanism for reducing blood glucose is insulin, which is a

N. Kuschinski (✉) · J. A. Christen
CIMAT, Guanajuato, Gto., Mexico
e-mail: nicolas.kuschinski@mat.uc.cl

A. Monroy
Hospital General, Mexico City, Mexico

S. Alavez
Universidad Autónoma Metropolitana, Unidad Lerma, Mexico City, Mexico

J. A. Christen et al. (eds.), *Statistics, Society and Environment*, Springer Proceedings in Mathematics & Statistics 479, https://doi.org/10.1007/978-3-031-78401-9_5

hormone produced by the pancreas in response to high levels of blood sugar. Diabetes occurs either because insulin production is insufficient or because the insulin being produced is ineffective. In either case, the result is that the body cannot reduce blood glucose to healthy levels [11].

For diagnosis of type 2 diabetes, one common test is the Oral Glucose Tolerance Test or OGTT. For this test, a fasting patient arrives, and his or her resting glucose is measured from a blood sample. The patient then drinks a 75 g glucose concentrate, and blood glucose is measured repeatedly over the course of the next 2 hours. The exact glucose measuring times vary depending on local practices. The results of these measurements are expected to provide some notion of how the patient's body handles the glucose [1, 4, 7].

In practice, the analysis of OGTT tests is usually done using very simple guidelines. Typically used markers include the average of the observed glucose measurements and/or the value of the first and last measurement. A patient is considered diabetic if the measurement chosen is above a certain threshold (typically 200 mg/dl). While this analysis has proven to be useful, it disregards one of the primary qualities of OGTT test conditions: that they measure the *evolution* of a process over time [4].

Accordingly, here we propose a dynamic model based on Ordinary Differential Equations (ODEs) to model blood glucose during an OGTT. The idea of using mathematical models to analyze OGTT results is not new. Previously proposed models have not been used for inference, mostly because they lack the flexibility to explain many of the phenomena seen in OGTTs. For instance, [7] assumes that the body only ever lowers blood glucose, but in the course of measuring several patients, we can see that this is not always the case (see real data in Sect. 4.1)

In our approach, we use a dynamic model, which was derived from the recommendations of our medical collaborators and follows the logic of previous related works such as [12]. It is flexible enough to describe most of the observed behavior of glucose in real patient's OGTTs. Using Bayesian inference, we put appropriate priors on the parameters and fit this model to real data. We have been able to achieve good fits for observed data, and our results match the intuition of our medical collaborators well enough that we consider it a good candidate to be considered for serious analysis of OGTT data and eventually for early diagnosis.

All of the data used for this chapter were obtained directly from Adriana Monroy (AM) from the Hospital General in Mexico City. It was all collected in the same location by the same team and for the same purpose.

This chapter is organized as follows: in Sect. 2, we present the dynamic model. In Sect. 3, we develop a Bayesian statistical model that can be used to draw inference from the dynamic model. In Sect. 4, we explain the details of how to perform inference from the model and present results of said inference on real patients. Finally, Sect. 5 concludes the chapter.

2 The Dynamic Model

Our model is based on the interaction of glucose, insulin, and glucagon only. The glucose regulation system is far more complex, but in the controlled environment of an OGTT, these are by far the leading factors [2]. Insulin is a hormone secreted by the pancreas, which reduces blood glucose. Glucagon is also a hormone produced in the pancreas and has the opposite effect; it triggers the liver to produce glucose, thus increasing blood glucose levels. In simple terms, insulin is produced when blood glucose is high, and glucagon is produced when bloRun!(Chain, T=1000, x0=start1, xp0=start2)od glucose is low, to make a feedback system of blood glucose level regulation [8, 12].

Our dynamical model is represented by the following system of ODEs:

$$\frac{dG}{dt} = L - I + \frac{D}{\theta_2} \tag{1}$$

$$\frac{dI}{dt} = \theta_0(G - G_b)^+ - \frac{I}{a} \tag{2}$$

$$\frac{dl}{dt} = \theta_1(G_b - G)^+ - \frac{L}{b} \tag{3}$$

$$\frac{dD}{dt} = -\frac{D}{\theta_2} + \frac{2V}{c} \tag{4}$$

$$\frac{dV}{dt} = -\frac{2V}{c} \tag{5}$$

The meaning of each of the state variables and parameters is explained in Table 1.

The heuristics behind this model are similar to other glucose–insulin models [12, for instance] and are as follows. There is a threshold level of glucose that the body hopes to maintain, which is denoted by G_b. It is set at 80 mg/dl for all

Table 1 Meanings of state variables and parameters in the OGTT model

	Interpretation	Value
G	Blood glucose	State variable
I	Blood Insulin	State variable
L	Blood Glucagon	State variable
D	Glucose in digestive system	State variable
V	Glucose not yet in the digestive system	State variable
θ_0	Insulin responsiveness	Unknown par.
θ_1	Glucagon responsiveness	Unknown par.
θ_2	Glucose digestive system mean life	Unknown par.
G_b	Base glucose level	Unknown par.
a, b	Insulin and Glucagon clearance mean life	31 min
c	Time taken to drink most of the glucose solution	5 min max

examples in this chapter, but it can be adjusted or inferred otherwise if that is deemed appropriate. If blood glucose goes above G_b, then insulin is produced, increasing $\frac{dI}{dt}$ as indicated by (2). As insulin is produced, this acts to reduce glucose concentration in the blood, reducing $\frac{dG}{dt}$ as indicated by (1). The opposite effect is achieved by glucagon, as seen in (3) and (1). Insulin and glucagon are both metabolized and decrease with mean lifes a and b as seen in Eqs. (2) and (3), respectively.

$D(t)$ and $V(t)$ represent glucose that is moving into the bloodstream. It begins outside the body, i.e., the sugar concentrate $V(t)$, decreasing and moving into the digestive system, $D(t)$, as seen in (5) and (4), and then from the digestive system moving into the bloodstream, as seen in (4) and (1).

Time is measured in hours, and blood glucose is measured in mg/dl of blood. The units of glucagon and insulin are more abstract and can be thought of in terms of their effect on units of blood glucose. Insulin and glucagon responsiveness includes both the generation of the hormone and the response of the body to the hormone after production. The model is not intended for insulin nor glucagon level prediction, and only glucose measurements are available, therefore in our model, the units of I and L are not relevant and not directly interpretable.

The values of a and b are extrapolated from best estimates of insulin and glucagon clearance time from [5]. Similarly, estimates exist on times of glucose absorption into the body [1], but these vary greatly from patient to patient and thus θ_2 is inferred. Jointly with θ_0 and θ_1, which are also inferred, these parameters represent the patient's condition in our model.

For the examples in this chapter, the system of ODEs is solved numerically (there is no known analytic solution). This is done by using the odeint function in the scipy package of the python programming language [9]. This uses an implementation of the LSODA algorithm, described in [13].

While some justification for the dynamic model comes from the heuristics, this is secondary to the real issue, which is whether its behavior can adequately represent what happens to glucose inside a patient's body. In Fig. 1, we see glucose curves, which follow from the model. The first three curves all start at G_b, to show the behavior of the model when the patient is already stable. One of the purposes of asking patients to fast beforehand is precisely to obtain this behavior—however, particularly for diabetic patients, fasting may not be sufficient, and glucose may begin elsewhere. The curves in the right panel of Fig. 1 represent a scenario wherein glucose begins somewhere other than G_b.

3 Statistical Model for the OGTT Data Analysis: The Inverse Problem

In order to perform inference on the OGTTs of real patients, the model must be fit to the data, i.e., the patient's glucose readings over the course of the test. For instance, for one real patient, at times $t = $ 0:00, 0:30, 1:00, 1:30, and 2:00

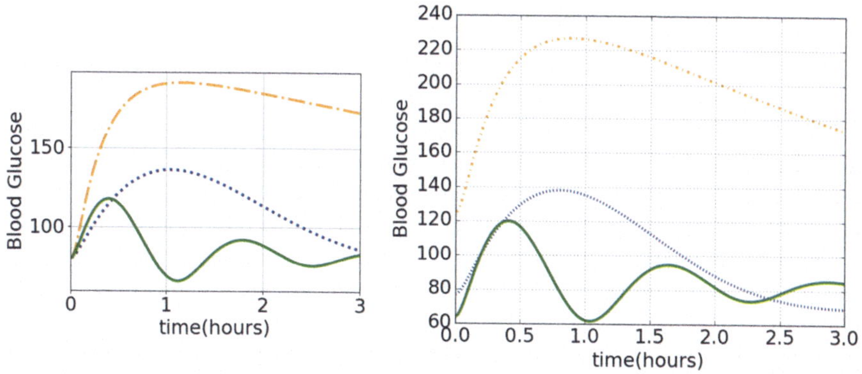

Fig. 1 **Left:** Three curves produced by our model, all beginning with $G(0) = G_b = 80\,\text{mg/dl}$; these represent three kinds of patients: The dotted line is a healthy normal patient, the broken line is a diabetic patient who does not adequately regulate insulin, and the solid line is an oscillating patient, whose insulin and glucagon response is very strong. **Right:** Curves showing similar scenarios but with slightly different parameter values, including $G(0)$

hours we obtained glucose measurements of $y = 81, 156, 141, 102$, and $89\,\text{mg/dl}$ respectively. The intent is to use these data to infer the glucose curves. We assume data to be observations of $G(t)$ at the measured times and model the data y with

$$y_i = G(t_i) + \epsilon_i$$

where $\epsilon_i \sim \mathcal{N}(0, \sigma^2)$ and $\sigma = 5\,\text{mg/dl}$ for all examples in this chapter. This allows us to write the likelihood as

$$f(y|\theta) \propto \prod_i e^{-\frac{(y_i - G(t_i|\theta))^2}{2\sigma^2}}.$$

Fitting this kind of model is considered an inverse problem; a nonlinear regression problem with a complex regressor defined through a system of ODEs. These systems are frequently characterized by drastically different sets of parameters fitting well with the same data, leading to many explanatory scenarios. For this reason, classical statistical estimators, such as the least squares estimator, only select one possible scenario, often quite an unreasonable one, following the data closely. There are several ways to address this issue, but one popular choice is to use Bayesian inference and encode some notion of what reasonable parameter combinations are in the prior distribution [6, 10].

For all of the examples in this chapter, the following priors were used:

$$\theta_0 \sim Gamma(2, 0.25)$$

$$\theta_1 \sim Gamma(2, 0.25)$$

$$\theta_2 \sim Gamma(10, 20)$$
$$G_b \sim \mathcal{N}(80, 20000) \text{ truncated to } [30, 400].$$

The priors for θ_0 and θ_1 were chosen to give high probability to all values estimated from even the most extreme patients that have been analyzed in this way. The prior for θ_2 is chosen to match information in [1]. The prior for G_b is centered on healthy patients and is truncated since any patient whose initial glucose is outside of this range should not undergo an OGTT test but instead be placed into emergency care [15] (AM performs a preliminary finger stick glucose test precisely for this purpose).

Within the range of reasonable values for the parameters, these priors are in general highly uninformative. This is important because the data collected for each patient are quite a small sample, and more informative priors would greatly affect inference. Similarly, it is a bad idea to use fully uninformative or reference priors. Estimating the parameters is an inference problem, and in inference problems, it is well known that highly unreasonable parameter combinations may result in regression functions that are very close to matching the data.

4 Inference

The object of interest is the function $G(t)$ for each patient, and inference is performed on data from each patient separately leading to a separate posterior for $\theta_0, \theta_1, \theta_2$ and G_b for each patient. Our objective here is not a population study, and hence, we concentrate on studying our model and its ability to fit OGTT data parsimoniously. Posterior exploration is achieved using MCMC techniques. Most MCMCs must be tuned to the posterior for each situation and in this case for each patient. A practical alternative is to use a self-tuning MCMC algorithm. One such algorithm is the t-walk, which is an MCMC algorithm that adapts to the scale of the target distribution. This is the algorithm that was chosen for this situation. It is a powerful MCMC algorithm, which can be used out of the box to simulate from a large family of target distributions. Further details on the mechanism by which the t-walk works can be found in [3].

Posterior exploration can be done in reasonable time even without high-end hardware. All the examples in this chapter were performed on a laptop computer with an i5 processor and 8 Gb of RAM. It took less than 2 minutes to perform 15,000 iterations of the t-walk. This represents 150 pseudo-independent posterior samples (using higher than necessary autocorrelation times, to account for patients with posterior distributions, which are harder to explore than usual). This is quite an acceptable numerical processing time, since it takes 2 hours to gather the blood samples, and processing is typically done overnight, depending on the availability of staff and laboratory equipment.

Convergence to stationarity of the chain is complex because it varies from patient to patient with each realization of the MCMC. We chose lengths for chains that were much longer than required to obtain the aforementioned 150 pseudo-independent samples for any observed data. For this chapter, we show posteriors based on 150 pseudo-independent samples, since these are sufficient for estimation of most general statistics, such as the mean or variance for each of the four parameters. If, in practice, finer detail is required, then longer chains can be produced in reasonable time.

4.1 Results

Figure 2 shows a posterior sample for three real patients whose OGTT curves were taken by AM at the Hospital General in Mexico City. The curves fit the data well, even for the third patient (top to bottom), whose data would not fit a curve, which does not account for glucagon. The first patient is a healthy patient, whose body handles glucose normally. The second patient is a potentially diabetic patient, whose glucose does not return to the baseline during the test. The third patient is a patient whose body responds rapidly to glucose, causing oscillations. Performing inference on many patients has shown that the latter is not an unusual or rare situation.

These curves display more nuance than current guidelines or practices for OGTT analysis. For instance, current practices would not distinguish between the first and third patients, despite their metabolism showing clearly different behavior, since the maximum measured value for both patients is similar (note also that although the first patient has higher glucose measurements, the third actually achieves a higher peak value in the estimated curve; this information can only be found by considering the temporal aspect of the measurements). Our model has strong descriptive power, giving reasonably small uncertainty for times in the measurement interval. It also has reasonable predictive power for a short time outside of the measurement interval as can be seen by prolonging the function $G(t)$ beyond the last measurement (in our graphs we prolong this an additional hour.) This can be thought of as a projection of what the patient's glucose *would* be if the conditions of the experiment were to continue. It is not clear, however, how long the dynamics of the system can be expected to remain intact, so this interpretation should only be considered over a short term.

Figures 3, 4, and 5 show histograms obtained from the MCMC posterior sampling for each of the model parameters for the patients from Fig. 2. The priors are represented with solid lines for reference. We may note that for patients without measurements below their resting glucose levels, the data are uninformative about θ_1, which represents glucagon response. This is to be expected since in our model, glucagon does not kick in unless blood glucose goes below G_b. Oscillating patients do provide data that are informative with regard to θ_1.

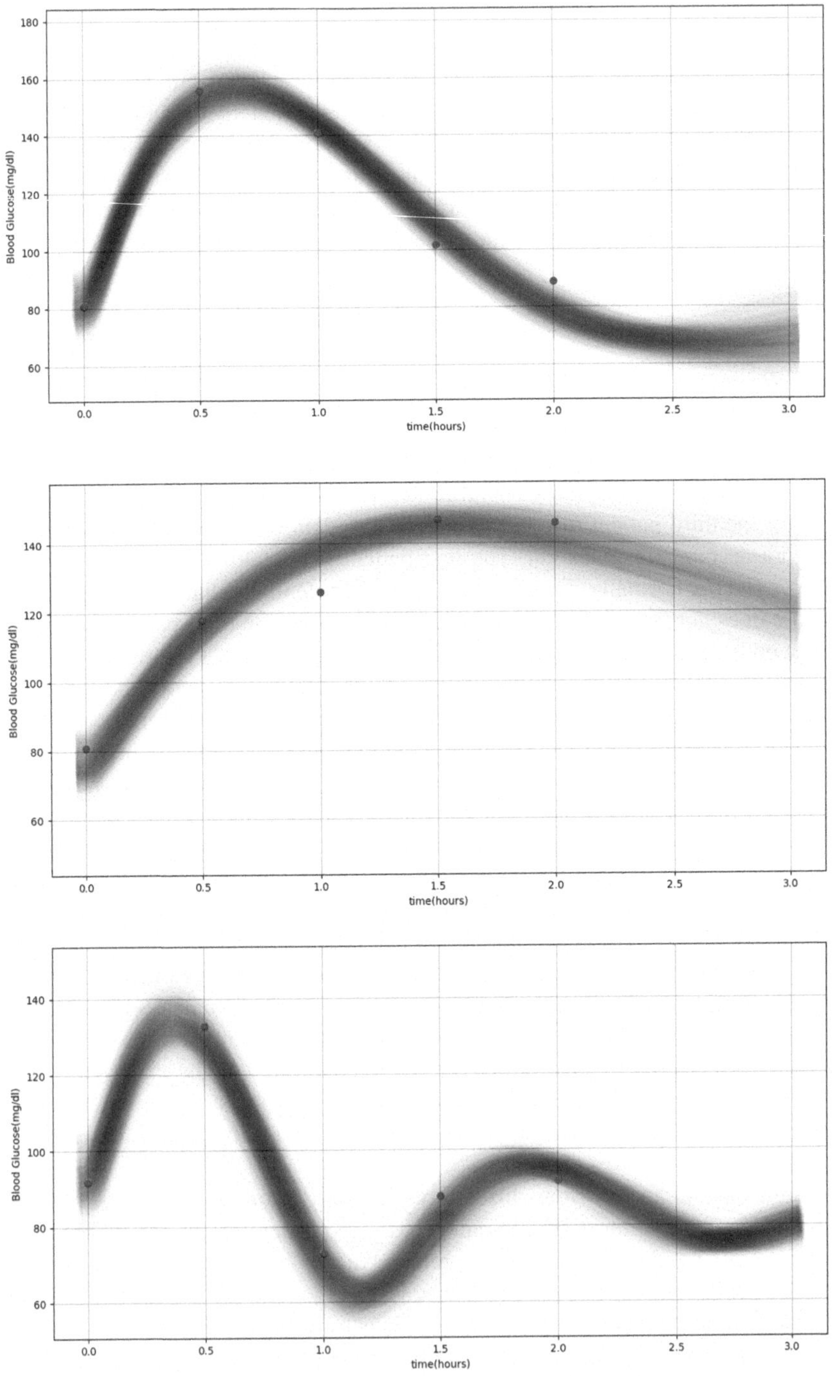

Fig. 2 OGTT inference for three patients. The first appears to be a healthy patient, the second a diabetic, and the third an oscillating case. The curves are pseudo-independent samples from the posterior distribution of $G(t)$ over 3 hours. The dots are the collected data. Colored lines indicate credible intervals and the Maximum A Posteriori estimates

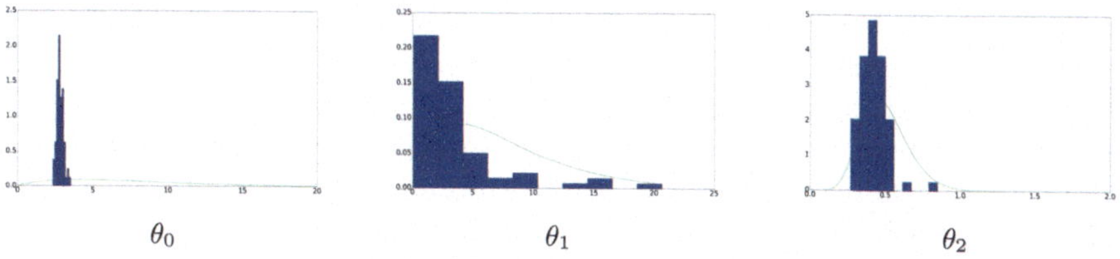

Fig. 3 Histograms of posterior samples for the first patient. They are the parameters θ_0, θ_1, and θ_2 respectively. They are superimposed on a graph of the prior density of each parameter. In particular, we note that for θ_1, the posterior matches the prior closely, and for θ_0, the data are extremely informative

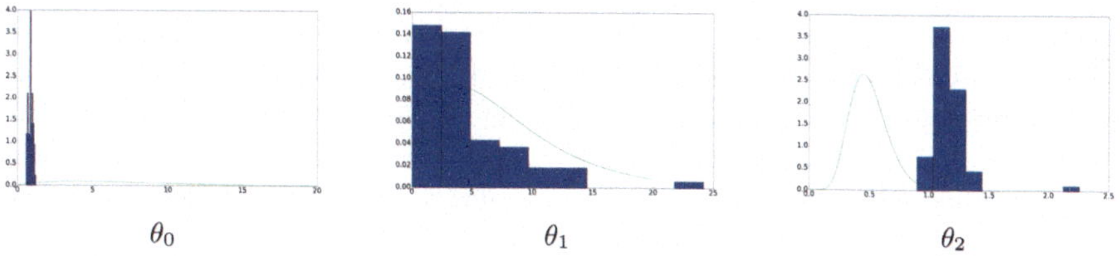

Fig. 4 Similar graphs for the second (potentially diabetic) patient. Once again we note that θ_1 deviates very little from the prior. This is caused by having no data below G_b

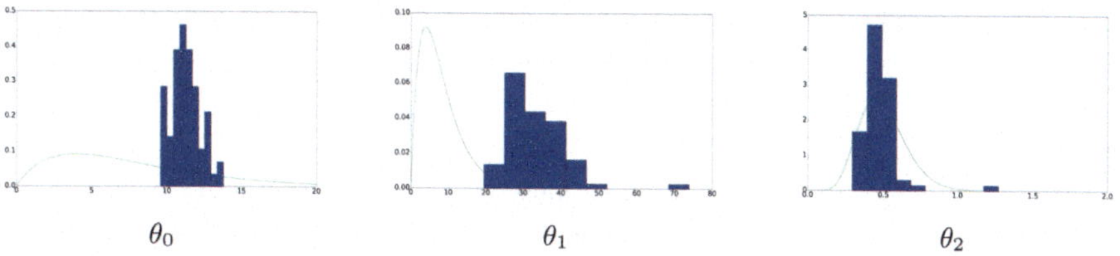

Fig. 5 Similar graphs for the third (oscillating) patient. We note that in this case, the data that are below G_b give us information about θ_1

5 Conclusions

The diagnosis of type 2 diabetes is an important public health issue, and it requires a more sophisticated tool than the direct recording of values from the test, not only because these values are insufficiently informative but also because they do not account for measurement error.

At present, our model has been used with data from over 100 patients, collected at various times by AM at the Hospital General in Mexico City. Of these patients, posterior inference was graphed and evaluated on whether it achieves an intuitively reasonable representation of the patient's curve, and whether this curve matched the clinical opinion of medical staff. Using five observations (measuring blood glucose every half hour), we failed to achieve seemingly good results for five patients. For some patients, a fit was attempted with three observations (measuring every hour). These data resulted in frequently multimodal distributions, which are difficult to interpret, and we believe that three observations are not enough to achieve good inference. AM obtained a sample of 16 patients for whom there are nine available measurements (measuring every 15 minutes). For these patients, all of the curves showed an intuitively good fit, which was a close match for the same data using only five of the available observations. Our model shows that overall it is able to represent the results of OGTT tests for nearly all patients for whom a fit was attempted. For one patient for whom the fit failed, it was later discovered that there was an error when recording the data, and the failure of the model to fit was an indication that triggered this error's discovery. The model also displays significantly deeper nuance and detail than previous analysis techniques could ever hope to represent.

Regarding the latter, a meta analysis of previously published studies could be attempted, to reconsider the conclusions reached in those reports. This may show light into, we envisage, two main points. Namely, to investigate the flexibility and applicability of our model on other OGTT data produced by other researchers and contrast the initial conclusions of those papers with those potentially suggested by our method. However, a meta analysis of this kind needs to be left for future research.

At present, this model serves for the analysis of OGTT data, but not for diagnosis. The reason for this is that this model provides much more information than had previously been available, and our medical collaborators are—as of yet—uncertain about how to interpret this new information for which they have not been trained. Further study is required in order to transform full glucose curves into diagnoses or treatment recommendations. That said, this kind of a study should be well worth the effort.

At present, there is no known method that serves to diagnose initial stages of type 2 diabetes quickly, and accurate diagnosis may only be done by following a patient over time. Regarding our model, we can envisage a faster and simpler solution based on a single-dimensional marker. One single-dimensional marker that seems reasonable is $min\left\{t : G(t) = G_b \text{ and } G'(t) < 0\right\}$ (first return of blood glucose to the base level G_b), although simply using the marginal posterior distribution for θ_0, and comparing it with references θ_0 values in healthy patients might also be a possibility.

We consider this model a strong candidate for further research in the analysis of OGTT data. However, even if not this specific model, some sort of dynamical model with strong descriptive power is required for the important and delicate tasks involved in the analysis of OGTT tests.

Acknowledgments JAC is partially funded by ONRG N62909-24-1-2016 grant.

References

1. Anderwald C, Gastaldelli A, Tura A, Krebs M, Promintzer-Schifferl M, Kautzky-Willer A, Stadler M, DeFronzo RA, Pacini G, Bischof MG (2011) Mechanism and effects of glucose

absorption during an oral glucose tolerance test among females and males. J Clin Endocrinol Metab 96(2):515–524
2. Carnevale Schianca GP, Rossi A, Sainaghi PP, Maduli E, Bartoli E (2003) The significance of impaired fasting glucose versus impaired glucose tolerance. Diabetes Care 26(5):1333–1337, URL https://doi.org/10.2337/diacare.26.5.1333
3. Christen JA, Fox C (2010) A general purpose sampling algorithm for continuous distributions (the t-walk). Bayesian Analysis 5(2):263–281, URL https://doi.org/10.1214/10-ba60
4. Davidson MB, Schriger DL, Peters AL, Lorber B (2000) Revisiting the oral glucose tolerance test criterion for the diagnosis of diabetes. Journal of General Internal Medicine 15:551–555, https://doi.org/10.1046/j.1525--1497.2000.08024.x
5. Duckworth WC, Bennett RG, Hamel FG (1998) Insulin degradation: Progress and potential. Endocrine Reviews 19(5):608–624, URL https://doi.org/10.1210/edrv.19.5.0349, pMID: 9793760
6. Fox C, Haario H, Christen J (2013) Inverse problems. In: Damien P, Dellaportas P, Polson N, Stephens D (eds) Bayesian Theory and Applications, Oxford University Press, chap 31, pp 619–643
7. Jansson L, Lindskog L, Nordén N, Carlström S, Scherstén B (1980) Diagnostic value of the oral glucose tolerance test evaluated with a mathematical model. Computers and Biomedical Research 13:512–521, https://doi.org/10.1016/0010-4809(80)90031-2
8. Jiang G, Zhang BB (2003) Glucagon and regulation of glucose metabolism. American Journal of Physiology-Endocrinology And Metabolism 284(4):E671–E678
9. Jones E, Oliphant T, Peterson P, et al. (2001–) SciPy: Open source scientific tools for Python. URL http://www.scipy.org/, [Online; accessed <today>]
10. Kaipio J, Somersalo E (2006) Statistical and computational inverse problems, vol 160. Springer Science & Business Media, New York
11. Metzger B (2006) American Medical Association guide to living with diabetes: preventing and treating type 2 diabetes: essential information you and your family need to know. John Wiley, Hoboken, N.J
12. Palumbo P, Ditlevsen S, Bertuzzi A, Gaetano AD (2013) Mathematical modeling of the glucose–insulin system: A review. Mathematical Biosciences 244(2):69–81, DOI https://doi.org/10.1016/j.mbs.2013.05.006, URL http://www.sciencedirect.com/science/article/pii/S0025556413001302
13. Petzold L (1983) Automatic selection of methods for solving stiff and nonstiff systems of ordinary differential equations. SIAM journal on scientific and statistical computing 4(1):136–148
14. Wild S, Roglic G, Green A, Sicree R, King H (2004) Global prevalence of diabetes: Estimates for the year 2000 and projections for 2030. http://care.diabetesjournals.org/content/27/5/1047.full
15. Yan JW, Azzam D, Columbus MP, Van Aarsen K, Liu SL, Spaic T, Shepherd LG (2021) Seeking care for hyperglycemia in the emergency department: Through the eyes of the patient. Diabetes Spectrum 34(1):60–66, https://doi.org/10.2337/ds20-0039

A Conditional Approach to Bayesian Inference in Copula Models

Arnoldo D. Miranda-Fournier, Joel Montesinos-Vázquez, and Gabriel Núñez-Antonio

Abstract An important problem in modeling real phenomena where several variables are involved is being able to have appropriate probabilistic multivariate models. On the other hand, it is also necessary to have methodologies that allow inferences to be made in these proposed models. From a probabilistic frame, a general approach to propose multivariate models is the use of copulas. However, being able to carry out a joint analysis for all parameters involved in copula models can become a challenging task. In this chapter, we propose an inference approach based on a decomposition of the final distribution. We show how to use that factorization to make joint Bayesian inferences for all parameters in models defined by a copula. The aforementioned approach offers the flexibility to carry out the inference process in two stages: a first stage in which inferences are made for parameters of marginal distributions involved in the corresponding copula, and a second step where inferences are obtained for parameters of the corresponding copula. The procedure is illustrated using simulated data.

Keywords Conditional sampling · Multivariate models · Bayesian inference

1 Introduction

A relevant problem in the analysis of phenomena involving several variables is to have multivariate models that adequately describe the behavior of these variables of interest. That description implies the understanding of dependency structure between the variables involved. In the literature, there are different ways to tackle the dependency problem between variables. However, the idea of copulas results in

A. D. Miranda-Fournier · J. Montesinos-Vázquez
Universidad Autónoma Metropolitana, Unidad Iztapalapa, Mexico City, México

G. Núñez-Antonio (✉)
Departamento de Matemáticas, Universidad Autónoma Metropolitana–Unidad Iztapalapa, Mexico City, México

J. A. Christen et al. (eds.), *Statistics, Society and Environment*, Springer Proceedings in Mathematics & Statistics 479, https://doi.org/10.1007/978-3-031-78401-9_6

a more general approach, since it allows establishing separately the behavior of the marginal distributions associated with the multivariate model and the dependency structure between the variables involved.

Although some ideas and results on copulas date back to the work of Hoeffding [7], the word *copula* appears until 1959 with the work of Sklar. The concept of copula is preceded by the technical definition of a multivariate probability distribution. However, here we will only approach its definition from an applied point of view. For a more technical survey, the reader is referred to Nelsen [9]. Thus, a copula can be defined as an m-variate probability distribution function on the unit hypercube $[0, 1]^m$, where the corresponding marginal functions are defined by uniform distributions on interval $[0, 1]$.

In situations where the joint final distribution of parameters of a copula model is difficult to analyze analytically, sampling methods such as Markov Chain Monte Carlo (MCMC) are relevant to make inferences for all parameters involved. Specifically, Smith [14] suggests the use of a Gibbs sampling to simulate from the respective joint posterior distribution. On the other hand, Silva and Lopes [12] propose a particular scheme in two stages to sample from the posterior distribution. See also, Pitt, Chan and Kohn [10], Ausin and Lopes [1], and Smith and Klein [15] for other approaches. However, it can often be complex to perform a joint sampling of the final distribution of all parameters in a copula model. The aforementioned difficulty is due, among other things, either to the dimensionality of parameter vector or to the structure of the corresponding complete conditional distributions, which simultaneously involve both parameters of marginal distributions as well as copula parameters. In this work, we present an decomposition of posterior distribution of all parameters involved in a copula model, which, if possible, allows to carry out inferences more efficiently.

This chapter is organized as follows. In Sect. 2, we present the definition of a copula for the multivariate case. In this section, we also show the Sklar's relevant theorem. In Sect. 3, we describe the proposed factorization based on a conditional approach. We also show how this approach can be applied to perform a joint sampling from the final distribution for all parameters in a model defined through copulas. Section 4 illustrates the proposed methodology with two simulated examples, where MCMC methods are used in conjunction with the conditional approach. Finally, in Sect. 5, we provide some final comments.

2 Copulas and the Sklar's Theorem

Copula theory allows us to solve the problem of defining or constructing multivariate distribution functions. This is done by associating the univariate marginal distributions with the joint distribution by means of the copula function. The latter thanks to Sklar's theorem. Before stating this result, we give the definition of copula.

A copula in $\mathbb{R}^m$ is a function

$$C : [0, 1]^m \rightarrow [0, 1]$$

that satisfies:

1. $C(u_1, u_2, \ldots, u_m) = 0$ if $u_j = 0$ for some $1 \leq j \leq m$.
2. $C(1, 1, \ldots, u_i, \ldots, 1) = u_i$, for $1 \leq i \leq m$.
3. For $\{a_1, \ldots, a_m\}$ and $\{b_1, \ldots, b_m\}$ such that $0 \leq a_i \leq b_i \leq 1$:

$$\sum_{i_1=1}^{2} \sum_{i_2=1}^{2} \cdots \sum_{i_m=1}^{2} (-1)^{i_1+\ldots+i_m} C(u_{1,i_1}, \ldots, u_{m,i_m}) \geq 0,$$

where $u_{j,1} = a_j$ and $u_{j,2} = b_j$ for $1 \leq j \leq m$.

In bivariate case, these three properties are equivalent to:

1. $C(u_1, 0) = C(0, u_2) = 0$.
2. $C(u_1, 1) = u_1$ and $C(1, u_2) = u_2$.
3. For $a_1 \leq b_1$ and $a_2 \leq b_2$:

$$C(b_1, b_2) + C(a_1, a_2) - C(b_1, a_2) - C(a_1, b_2) \geq 0.$$

The importance of a copula can be understood through Sklar's theorem.
Sklar's Theorem ([13]) If $F(X_1, \ldots, X_m)$ is a distribution function with univariate marginals $F_1(X_1), \ldots, F_m(X_m)$, then there is a copula C such that

$$F(X_1, \ldots, X_m) = C(F_1(X_1), \ldots, F_m(X_m)). \tag{1}$$

Reciprocally, if C is a copula and $F_1(X_1), \ldots, F_m(X_m)$ are univariate distribution functions, then

$$C(F_1(X_1), \ldots, F_m(X_m)) \tag{2}$$

is a m-variate distribution function. ■

The previous result establishes the relationship between the joint distribution and the corresponding univariate marginals, as well as a way to construct multivariate distributions. A proof of Sklar's theorem can be found in [9].

On the other hand, if we define the multivariate distribution function as in the Eq. (2), then the probability density function (pdf) is

$$f(x_1, \ldots, x_m) = c(F_1(x_1), \ldots, F_m(x_m))\, f_1(x_1) \cdots f_m(x_m), \tag{3}$$

where f_i are the pdf for each x_i and c is the derivative of C given by

$$c(u_1, \ldots, u_m) = \frac{\partial^m}{\partial u_1 \cdots \partial u_m} C(u_1, \ldots, u_m). \tag{4}$$

According to the way copulas are constructed, these can be classified as Archimedean (Frank, Clayton, Gumbel), Implicit (Gaussian, Student's t), Vine, or Empirical.

2.1 *Implicit Copulas*

In this work, we focus on implicit copulas. These are constructed from multivariate densities, following the inversion method [9]. Thus, if H is a m-variate distribution function with univariate marginals $H_1, \ldots, H_m$, then

$$C_H(u_1, \ldots, u_m) = H(H_1^{-1}(u_1), \ldots, H_m^{-1}(u_m)) \tag{5}$$

is a m-variate copula, and the corresponding copula density is given by

$$c_H(u_1, \ldots, u_m) = \frac{h(H_1^{-1}(u_1), \ldots, H_m^{-1}(u_m))}{h_1(H_1^{-1}(u_1)) \cdots h_m(H_m^{-1}(u_m))}, \tag{6}$$

where h and h_i are the pdf associated to H and to each H_i, respectively. Thus, setting a model H (which determines an implicit copula C_H), a distribution function m-variate F with marginals $F_1(x_1), \ldots, F_m(x_m)$ can be defined as follows:

$$\begin{aligned} F(x_1, \ldots, x_m) &= C_H(F_1(x_1), \ldots, F_m(x_m)) \\ &= H(H_1^{-1}(F_1(x_1)), \ldots, H_m^{-1}(F_m(x_m))). \end{aligned} \tag{7}$$

If the marginal density functions for every x_i are denoted by f_i, $i = 1, 2, \ldots, m$, then the density function associated with F can be written as

$$\begin{aligned} f(x_1, \ldots, x_m) &= c_H(F_1(X_1), \ldots, F_m(X_m)) \cdot f_1(x_1) \cdots f_m(x_m) \\ &= \frac{h(H_1^{-1}(F_1(x_1)), \ldots, H_m^{-1}(F_m(x_m)))}{h_1(H_1^{-1}(F_1(x_1))) \cdots h_m(H_m^{-1}(F_m(x_m)))} f_1(x_1) \cdots f_m(x_m). \end{aligned}$$

Next, we specialize the discussion above for the t-copula as well as the Gaussian copula.

2.1.1 The t-Copula

In this case $h(\cdot) = T_m(\cdot \mid \mathbf{0}, \mathbf{R}, \nu)$, where T_m is the distribution function of a m-variate Student's t variable with correlation matrix $\mathbf{R}$, ν degrees of freedom ($\nu > 2$), and mean $\mathbf{0}$. From Eq. (7), we obtain that the t-copula with parameters $\mathbf{R}$ and ν is given by

$$C_t(u_1, \ldots, u_m \mid \mathbf{R}, \nu) = T_m(T_\nu^{-1}(u_1), \ldots, T_\nu^{-1}(u_m) \mid \mathbf{0}, \mathbf{R}, \nu), \tag{8}$$

where T_ν is the distribution function of a univariate variable Student's t with ν degrees of freedom. The corresponding copula density is

$$c_t(u_1, \ldots, u_m \mid \mathbf{R}, \nu) = \frac{t_m(T_\nu^{-1}(u_1), \ldots, T_\nu^{-1}(u_m) \mid \mathbf{0}, \mathbf{R}, \nu)}{t_\nu(T_\nu^{-1}(u_1)) \cdots t_\nu(T_\nu^{-1}(u_m))}, \tag{9}$$

where $t_m(\cdot \mid \mathbf{0}, \mathbf{R}, \nu)$ denotes the density function of a m-variate Student's t and $t_\nu(\cdot)$ the density of a univariate Student's t with ν degrees of freedom.

In the bivariate case ($m = 2$), the correlation matrix $\mathbf{R}$ is determined by a correlation coefficient ρ,

$$\mathbf{R} = \begin{bmatrix} 1 & \rho \\ \rho & 1 \end{bmatrix}. \tag{10}$$

Therefore, the bivariate t-copula with parameters $\nu \in (2, \infty)$ and $\rho \in [-1, 1]$ is expressed as

$$C_t(u_1, u_2 \mid \rho, \nu) = T_2(T_\nu^{-1}(u_1), T_\nu^{-1}(u_2) \mid \mathbf{0}, \rho, \nu). \tag{11}$$

2.1.2 The Gaussian Copula

If $h(\cdot) = F_m(\cdot \mid \mathbf{R})$ denotes the distribution function of the m-variate normal $N_m(\mathbf{0}, \mathbf{R})$, with mean $\mathbf{0}$ and $\mathbf{R}$ a correlation matrix of dimension $m \times m$, then the Gaussian copula with parameter $\mathbf{R}$ is defined as

$$C_G(u_1, \ldots, u_m \mid \mathbf{R}) = F_m(\Phi^{-1}(u_1), \ldots, \Phi^{-1}(u_m) \mid \mathbf{R}), \tag{12}$$

where Φ is the distribution function of a univariate standard normal. The corresponding copula density is given by

$$c_G(u_1, \ldots, u_m \mid \mathbf{R}) = \frac{f_m(\Phi^{-1}(u_1), \ldots, \Phi^{-1}(u_m) \mid \mathbf{R})}{\phi(\Phi^{-1}(u_1)) \cdots \phi(\Phi^{-1}(u_m))}, \tag{13}$$

where $f_m(\cdot|\mathbf{R})$ denotes the density function of $N_m(\mathbf{0}, \mathbf{R})$ and $\phi(\cdot)$ the density of a univariate standard normal. As in the case of the t-copula, when $m = 2$, the correlation matrix $\mathbf{R}$ is completely determined from a correlation coefficient ρ (Eq. 10). Therefore, in the bivariate case, the Gaussian copula with parameter ρ is expressed as

$$C_G(u_1, u_2|\rho) = F_2(\Phi^{-1}(u_1), \Phi^{-1}(u_2) \mid \rho). \tag{14}$$

It is important to emphasize that copulas have an invariance property under strictly increasing transformations [9]. That implies both the Gaussian copula and the t-copula remain invariant under the standardization of their marginal distributions [4]. In the case of the Gaussian copula, the copula generated through the normal $N_p(\boldsymbol{\mu}, \boldsymbol{\Sigma})$ is identical to that generated through the normal $N_p(\mathbf{0}, \mathbf{R})$, where $\mathbf{R}$ is the implicit correlation matrix in the variance and covariance matrix $\boldsymbol{\Sigma}$. Similarly, the copula generated by the multivariate Student's t $T_p(\boldsymbol{\mu}, \boldsymbol{\Sigma}, \nu)$ is identical to that generated by $T_p(\mathbf{0}, \mathbf{R}, \nu)$, with $\mathbf{R}$ being the correlation matrix associated with the variance and covariance matrix $\boldsymbol{\Sigma}$.

2.2 Sampling from Implicit Copulas

An important aspect in the analysis of the performance of inference processes in copula models is the ability to simulate samples of the corresponding copula model. In this section, we will show how to obtain a sample from a implicit copula, defined through a multivariate distribution H and its marginals H_i.

1. Generate a sample of size n from the m-variate distribution H,

$$V^{(1)}, V^{(2)}, \ldots, V^{(n)} \sim H. \tag{15}$$

It should be noted that each $V^{(j)}$ has the form $V^{(j)} = (v_{1j}, v_{2j}, \ldots, v_{mj})$.

2. For each $j = 1, \ldots, n$, construct the variable $U^{(j)}$ as

$$\begin{aligned} U^{(j)} &= (H_1(v_{1j}), H_2(v_{2j}), \ldots, H_m(v_{mj})) \\ &= (u_{1j}, u_{2j}, \ldots, u_{mj}). \end{aligned} \tag{16}$$

In this way, $U^{(1)}, U^{(2)}, \ldots, U^{(n)}$ is a sample from the copula C_H.

In order to understand the aforementioned procedure, we present the following ideas. It can be noted that for i fixed, $\{u_{i1}, u_{i2}, \ldots, u_{in}\}$ is a sample of a Uniform random variable, which we denote as U_i; and $\{v_{i1}, v_{i2}, \ldots, v_{in}\}$ is a sample from the marginal V_i with distribution H_i. Then

$$\begin{aligned} \mathbb{P}[U_1 < u_1, \ldots, U_m < u_m] &= \mathbb{P}\left[H_1^{-1}(U_1) < H_1^{-1}(u_1), \ldots, H_m^{-1}(U_m) < H_m^{-1}(u_m)\right] \\ &= \mathbb{P}\left[V_1 < H_1^{-1}(u_1), \ldots, V_m < H_m^{-1}(u_m)\right] \\ &= H\left(H_1^{-1}(u_1), \ldots, H_m^{-1}(u_m)\right) \\ &= C_H(u_1, \ldots, u_m). \end{aligned} \tag{17}$$

As a particular example of the aforementioned algorithm, here we show the bivariate case considering the Gaussian copula $C_G(u_1, u_2|\rho)$, where $H = N_2(\mathbf{0}, \boldsymbol{R})$. Here, we take $\boldsymbol{R}$ as in Eq. (10). Thus, to generate a sample $\{U^{(i)}\}$, $i = 1, \ldots, n$ of size n from the copula $C_G(u_1, u_2|\rho)$, the algorithm would be as follows:

1. Sampling $V^{(i)} = (v_{1i}, v_{2i})$ from a distribution $N_2(\mathbf{0}, \boldsymbol{R})$, for $i = 1, \ldots, n$.
2. For each $i = 1, \ldots, n$, take $U^{(i)}$ as

$$\begin{aligned} U^{(i)} =& (\Phi(v_{1i}), \Phi(v_{2i})) \\ =& (u_{1i}, u_{2i}), \end{aligned}$$

where Φ is the distribution function of a standard univariate normal variable.

In this way, the vectors $U^{(i)}$, $i = 1, \ldots, n$ constitute a sample of size n from the copula $C_G(u_1, u_2 \mid \rho)$.

3 Conditional Approach for Bayesian Inference

For a better presentation of ideas, in this section, we specialize the discussion to the bivariate case, the generalization to the multivariate case follows naturally. Let $X = (X_1, X_2)$ be a bivariate random vector with density function given by

$$\begin{aligned} f(x_1, x_2 \mid \omega_1, \omega_2, \omega_C) = & \; c(F_1(x_1 \mid \omega_1), F_2(x_2 \mid \omega_2) \mid \omega_C) \\ & \times f_1(x_1 \mid \omega_1)\, f_2(x_2 \mid \omega_2), \end{aligned}$$

where $\boldsymbol{\omega}_1, \boldsymbol{\omega}_2$ are the marginal parameter vectors of X_1 and X_2 respectively, and $\boldsymbol{\omega}_C$ represents the parameter vector of copula density function $c(\cdot, \cdot \mid \cdot)$.

If $\boldsymbol{\omega}_X = (\boldsymbol{\omega}_1, \boldsymbol{\omega}_2)$ and $\boldsymbol{\omega}_C$ are unknown, then we want to make joint Bayesian inferences for them. Thus, if $\pi(\boldsymbol{\omega}_X, \boldsymbol{\omega}_C)$ is a prior distribution for these parameters and $\boldsymbol{D}_n = \{(x_{1i}, x_{2i})\}_{i=1,\ldots,n}$ is a sample of size n from $f(x_1, x_2 \mid \boldsymbol{\omega}_1, \boldsymbol{\omega}_2, \boldsymbol{\omega}_C)$, then the posterior distribution for $(\boldsymbol{\omega}_X, \boldsymbol{\omega}_C)$ is given by

$$\pi(\boldsymbol{\omega}_X, \boldsymbol{\omega}_C \mid \boldsymbol{D}_n) = \frac{\prod_{i=1}^n f(x_{1i}, x_{2i} \mid \boldsymbol{\omega}_X, \boldsymbol{\omega}_C)\, \pi(\boldsymbol{\omega}_X, \boldsymbol{\omega}_C)}{\int \prod_{i=1}^n f(x_{1i}, x_{2i} \mid \boldsymbol{\omega}_X, \boldsymbol{\omega}_C)\, \pi(\boldsymbol{\omega}_X, \boldsymbol{\omega}_C)\, d\boldsymbol{\omega}_X\, d\boldsymbol{\omega}_C}. \quad (18)$$

If the posterior distribution (18) is complicated to analyze analytically, an option for making inferences for $(\boldsymbol{\omega}_X, \boldsymbol{\omega}_C)$ is sampling from the posterior density $\pi(\boldsymbol{\omega}_X, \boldsymbol{\omega}_C \mid \boldsymbol{D}_n)$. That leads us to the problem of obtaining a sample from a multivariate distribution. Different methods to solve that problem have been proposed in the literature. In particular [14] suggests a Gibbs sampling using the conditionals

$\pi(\boldsymbol{\omega}_X \mid \boldsymbol{\omega}_C, \boldsymbol{D}_n)$ and $\pi(\boldsymbol{\omega}_C \mid \boldsymbol{\omega}_X, \boldsymbol{D}_n)$. On the other hand, Silva and Lopes [12] propose to make the estimates in two stages, in the first use the mode $m_{\omega X}$ of $\pi(\boldsymbol{\omega}_X \mid \boldsymbol{D}_n)$ and subsequently sample from $\pi(\boldsymbol{\omega}_C \mid \boldsymbol{\omega}_X = m_{\omega_X}, \boldsymbol{D}_n)$. See also [1, 10, 15] for others approaches. The proposal in this work is to construct a decomposition of the posterior distribution (18), as follows:

$$\pi(\boldsymbol{\omega}_X, \boldsymbol{\omega}_C \mid \boldsymbol{D}_n) = \pi(\boldsymbol{\omega}_X \mid \boldsymbol{D}_n)\, \pi(\boldsymbol{\omega}_C \mid \boldsymbol{\omega}_X, \boldsymbol{D}_n). \tag{19}$$

In this way, we can use the conditional method of Rosenblatt [11] (see also [8]), to generate a sample of the multivariate density (18). It is relevant to mention that in order to implement this approach, which we will call *conditional approach*, one must be able to identify the conditional density $\pi(\boldsymbol{\omega}_C \mid \boldsymbol{\omega}_X, \boldsymbol{D}_n)$, as well as the marginal density $\pi(\boldsymbol{\omega}_X \mid \boldsymbol{D}_n)$. If the decomposition (19) is feasible, then a sampling from the posterior distribution $\pi(\boldsymbol{\omega}_X, \boldsymbol{\omega}_C \mid \boldsymbol{D}_n)$, using the conditional approach, can be implemented as:

1. Sampling $\boldsymbol{w}_X^*$ from the marginal distribution $\pi(\boldsymbol{\omega}_X \mid \boldsymbol{D}_n)$.
2. Sampling $\boldsymbol{w}_C^*$ from the conditional distribution $\pi(\boldsymbol{\omega}_C \mid \boldsymbol{\omega}_X = \boldsymbol{w}_X^*, \boldsymbol{D}_n)$.

Thus, using the aforementioned conditional approach, the sampling problem from the posterior distribution $\pi(\boldsymbol{\omega}_X, \boldsymbol{\omega}_C \mid \boldsymbol{D}_n)$ is reduced to identify and sample from the posterior densities $\pi(\boldsymbol{\omega}_X \mid \boldsymbol{D}_n)$ and $\pi(\boldsymbol{\omega}_C \mid \boldsymbol{\omega}_X, \boldsymbol{D}_n)$. Thus, the first problem is to be able for obtaining the factorization (19). In general, this step can be attacked as follows:

$$\pi(\boldsymbol{\omega}_X \mid \boldsymbol{D}_n) = \int \pi(\boldsymbol{\omega}_X, \boldsymbol{\omega}_C \mid \boldsymbol{D}_n)\, d\boldsymbol{\omega}_C,$$

and

$$\pi(\boldsymbol{\omega}_C \mid \boldsymbol{\omega}_X, \boldsymbol{D}_n) = \frac{\pi(\boldsymbol{\omega}_X, \boldsymbol{\omega}_C \mid \boldsymbol{D}_n)}{\pi(\boldsymbol{\omega}_X \mid \boldsymbol{D}_n)}.$$

However, the aforementioned integral can be complicated and cannot always be solved analytically. Fortunately, in order to implement most MCMC sampling methods, it is sufficient to obtain expressions proportional to $\pi(\boldsymbol{\omega}_X \mid \boldsymbol{D}_n)$ and $\pi(\boldsymbol{\omega}_C \mid \boldsymbol{\omega}_X, \boldsymbol{D}_n)$. In this regard, the following result is useful for obtaining such proportional expressions (see Appendix for the corresponding proof).

Theorem 1 *Let f be a function on $\mathbb{R}^m$. Suppose that f can be written as*

$$f(X_1, X_2) \propto g_1(X_1, X_2)\, g_2(X_1),$$

where $g_2 : \mathbb{R}^l \to \mathbb{R}$ only depends on $X_1 \in \mathbb{R}^l$ and $g_1 : \mathbb{R}^m \to \mathbb{R}$. Then

$$f(X_2 \mid X_1 = x_1) \propto g_1(x_1, X_2),$$

and

$$f(X_1) \propto g_2(X_1),$$

where $f(X_2 \mid X_1 = x_1)$ *is the conditional density of* $X_2 \mid X_1$ *and* $f(X_1)$ *is the marginal density of* X_1. ■

Thus, if the prior $\pi(\boldsymbol{\omega}_X, \boldsymbol{\omega}_C)$ has an independent structure or the form $\pi(\boldsymbol{\omega}_C \mid \boldsymbol{\omega}_X)\pi(\boldsymbol{\omega}_X)$, then

$$\begin{aligned}\pi(\boldsymbol{\omega}_X, \boldsymbol{\omega}_C \mid \boldsymbol{D}_n) &\propto \prod_{i=1}^{n} c(F(x_{1i} \mid \boldsymbol{\omega}_1), F(x_{2i} \mid \boldsymbol{\omega}_1) \mid \boldsymbol{\omega}_C) \\ &\quad \times f(x_{1i} \mid \boldsymbol{\omega}_1)\, f(x_{2i} \mid \boldsymbol{\omega}_2)\, \pi(\boldsymbol{\omega}_X)\, \pi(\boldsymbol{\omega}_C) \\ &= g_1(\boldsymbol{\omega}_X, \boldsymbol{\omega}_C) \times g_2(\boldsymbol{\omega}_X),\end{aligned}$$

where

$$g_1(\boldsymbol{\omega}_X, \boldsymbol{\omega}_C) = \prod_{i=1}^{n} c(F(x_{1i} \mid \boldsymbol{\omega}_1), F(x_{2i} \mid \boldsymbol{\omega}_2) \mid \boldsymbol{\omega}_C)\, \pi(\boldsymbol{\omega}_C),$$

and

$$g_2(\boldsymbol{\omega}_X) = \prod_{i=1}^{n} f(x_{1i} \mid \boldsymbol{\omega}_1) f(x_{2i} \mid \boldsymbol{\omega}_2)\, \pi(\boldsymbol{\omega}_X).$$

By the Theorem 1 it follows that

$$\pi(\boldsymbol{\omega}_C \mid \boldsymbol{\omega}_X, \boldsymbol{D}_n) \propto \prod_{i=1}^{n} c(F(x_{1i} \mid \boldsymbol{\omega}_1), F(x_{2i} \mid \boldsymbol{\omega}_1) \mid \boldsymbol{\omega}_C)\, \pi(\boldsymbol{\omega}_C), \tag{20}$$

and

$$\pi(\boldsymbol{\omega}_X \mid \boldsymbol{D}_n) \propto \prod_{i=1}^{n} f(x_{1i} \mid \boldsymbol{\omega}_1) f(x_{2i} \mid \boldsymbol{\omega}_2)\, \pi(\boldsymbol{\omega}_X). \tag{21}$$

The expressions given by the Eqs. (20) and (21) are indispensable to perform the conditional approach and thus be able to sample efficiently from the posterior density $\pi(\boldsymbol{\omega}_X, \boldsymbol{\omega}_C \mid \boldsymbol{D}_n)$.

4 Examples

4.1 Example 1

In this example, we consider a 3-variate model defined by a Gaussian copula with marginals gamma distribution. Specifically, the model is defined as

$$f(x, y, z \mid \omega_1, \omega_2, \omega_3, \omega_C) = c_G(F_X(x \mid \omega_1), F_Y(y \mid \omega_2), F_Z(z \mid \omega_3) \mid \omega_C) \\ \times Ga(x \mid \omega_1)\, Ga(y \mid \omega_2)\, Ga(z \mid \omega_3).$$

where $c_G(\cdot \mid \cdot)$ is the copula density defined in (13) with $m = 3$, and $F_R(r|\omega_j)$ is the cumulative distribution function of a Gamma variable R with parameters $\omega_j = (\alpha_j, \beta_j)$ for $j = 1, 2, 3$. Here, the corresponding density is given by

$$Ga(r|\alpha, \beta) = \frac{\beta^\alpha}{\Gamma(\alpha)} r^{\alpha-1} e^{-\beta r}.$$

We simulated a sample $\boldsymbol{D}_n = \{(x_1, y_1, z_1), \ldots (x_n, y_n, z_n)\}$ of size $n = 1000$ from the model $f(x, y, z \mid \omega_1, \omega_2, \omega_3, \omega_C)$ where $\omega_1 = (5, 4)$, $\omega_2 = (2, 1.5)$, $\omega_3 = (1, 2)$ and $\omega_C = (-0.9, 0.9, -0.9)$. Figure 1 shows the simulated data.

If $\pi(\omega_1, \omega_2, \omega_3, \omega_c)$ represents the prior distribution for model parameters, then the posterior distribution up to a constant of proportionality is given by

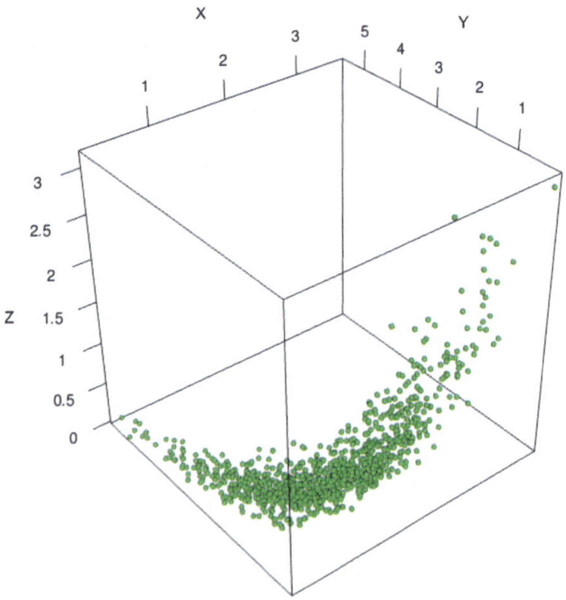

Fig. 1 Sample $\{(x_i, y_i, z_i)\}_{1 \le i \le n}$ of size $n = 1000$ from model $f(x, y, z \mid \omega_1, \omega_2, \omega_3, \omega_C)$ of Example 1

$$\begin{aligned}\pi(\omega_1, \omega_2, \omega_3, \omega_C \mid \boldsymbol{D}_n) \propto\ & f(\boldsymbol{D}_n \mid \omega_1, \omega_2, \omega_3, \omega_C)\, \pi(\omega_1, \omega_2, \omega_3, \omega_C) \\ =\ & \prod_{i=1}^{n} c_G(F_X(x_i \mid \omega_1), F_Y(y_i \mid \omega_2), F_Z(z_i \mid \omega_3) \mid \omega_C) \\ & \times Ga(x_i \mid \omega_1)\, Ga(y_i \mid \omega_2)\, Ga(z_i \mid \omega_3) \\ & \times \pi(\omega_1, \omega_2, \omega_3, \omega_C). \end{aligned} \tag{22}$$

For this example, we use the next prior specification:

$$\begin{aligned}\pi(\omega_1, \omega_2, \omega_3, \omega_C) =\ & U(\rho_{12} \mid -1, 1)\, U(\rho_{13} \mid -1, 1)\, U(\rho_{23} \mid -1, 1) \\ & \prod_{j=1}^{3} Ga(\alpha_j \mid a_j, b_j)\, Ga(\beta_j \mid c_j, d_j), \end{aligned} \tag{23}$$

where $a_j = c_j = 1, b_j = d_j = 0.001$ for $j = 1, 2, 3$.

Thus, the posterior (22) can be expressed as

$$\begin{aligned}\pi(\omega_1, \omega_2, \omega_3, \omega_C \mid \boldsymbol{D}_n) =\ & \prod_{i=1}^{n} c_G(F(x_i \mid \omega_1), F(y_i \mid \omega_2), F(z_i \mid \omega_3) \mid \omega_C) \\ & \times U(\rho_{12} \mid -1, 1)\, U(\rho_{13} \mid -1, 1)\, U(\rho_{23} \mid -1, 1) \\ & \times \prod_{i=1}^{n} Ga(x_i \mid \omega_1)\, Ga(y_i \mid \omega_2)\, Ga(z_i \mid \omega_3) \\ & \times \prod_{j=1}^{3} Ga(\alpha_j \mid a_j, b_j)\, Ga(\beta_j \mid c_j, d_j) \qquad (24) \\ =\ & \pi(\omega_C \mid \omega_1, \omega_2, \omega_3, \boldsymbol{D}_n)\, \pi(\omega_1, \omega_2, \omega_3 \mid \boldsymbol{D}_n). \end{aligned}$$

We use the conditional approach described in previous section for sampling from the posterior (24). First, we implemented *Adaptative Rejection Sampling* (ARS) algorithms within a Gibbs sampler for sampling from distribution $\pi(\omega_1, \omega_2, \omega_3 \mid \boldsymbol{D}_n)$. For details about that algorithm, the reader is referred to Gilks and Wild [5], Gilks, et al. [6], and Biswabrata and Debasis [2]. After, we use a Gibbs sampler for sampling from distribution $\pi(\omega_C \mid \omega_1, \omega_2, \omega_3, \boldsymbol{D}_n)$. The latter, through the complete conditional densities $\pi(\rho_{12} \mid \rho_{13}, \rho_{23}, \omega_1, \omega_2, \omega_3, \boldsymbol{D}_n)$, $\pi(\rho_{13} \mid \rho_{12}, \rho_{23}, \omega_1, \omega_2, \omega_3, \boldsymbol{D}_n)$ and $\pi(\rho_{23} \mid \rho_{12}, \rho_{13}, \omega_1, \omega_2, \omega_3, \boldsymbol{D}_n)$.

Columns 1–3 of the Table 1 show the 95% posterior credible intervals for each elements of the vectors ω_1, ω_2, ω_3 and ω_C. It can be seen that the true value of each of the parameters is well inside the corresponding credible interval. On the other hand, as a way to evaluate the sensitivity of the obtained inferences, with respect to the specification of the initial distributions used, we apply the proposed methodology using other non-informative prior specifications. Columns 4 and 5 of Table 1 show the corresponding credibility intervals. In particular, it can be

Table 1 95% credible intervals for each parameters of vectors ω_1, ω_2, ω_3, and ω_C of the model of Example 1

Parameter	Real value	Credibility intervals		
		$b_j = d_j = 0.001$	$b_j = d_j = 0.01$	$b_j = d_j = 0.0001$
α_1	5	(4.80, 5.62)	(4.72, 5.63)	(4.72, 5.62)
α_2	2	(1.95, 2.29)	(1.96, 2.30)	(1.95, 2.30)
α_3	1	(0.96, 1.12)	(0.96, 1.11)	(0.95, 1.11)
β_1	4	(3.80, 4.51)	(3.76, 4.51)	(3.76, 4.49)
β_2	1.5	(1.47, 1.76)	(1.47, 1.77)	(1.45, 1.75)
β_3	2	(1.89, 2.30)	(1.92, 2.31)	(1.89, 2.30)
ρ_{12}	−0.9	(−0.91, −0.89)	(−0.91, −0.89)	(−0.91, −0.88)
ρ_{13}	0.9	(0.88, 0.91)	(0.88, 0.91)	(0.88, 0.91)
ρ_{23}	−0.9	(−0.91, −0.88)	(−0.91, −0.88)	(−0.91, −0.88)

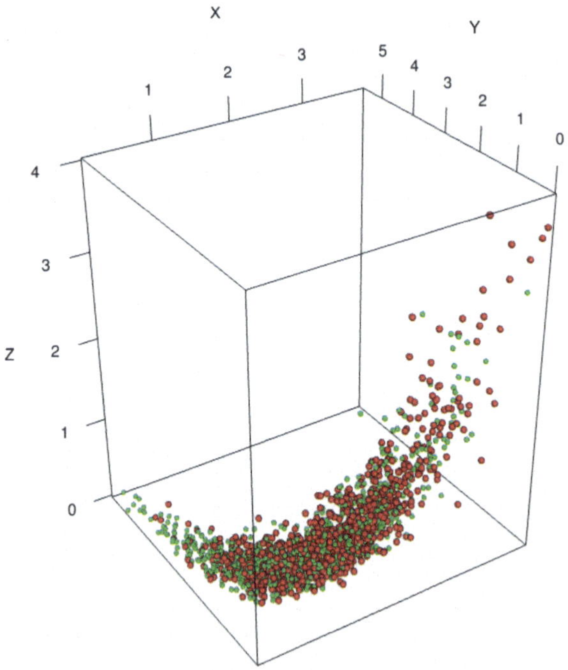

Fig. 2 Green points: sample $\{(x_i, y_i, z_i)\}_{1\le i\le n}$ from model $f(x, y, z \mid \omega_1, \omega_2, \omega_3, \omega_C)$. Red points: Sample from the corresponding predictive distribution

observed that the inferences obtained are not sensitive to the specification of the non-informative prior distributions. Finally, Fig. 2 shows the corresponding predictive distribution (red points), obtained using the methodology proposed in this work.

It should be mentioned that a raw Gibbs sampling could have been implemented using the corresponding completed conditional densities for all parameters $\alpha_1, \alpha_2, \alpha_3, \beta_1, \beta_2, \beta_3, \rho_{12}, \rho_{13}$ and ρ_{23}. However, as we will see in the following example, that *joint* procedure is more expensive in terms of convergence and therefore requires more computational effort.

4.2 Example 2

In this example, we consider a bivariate model defined by a t-copula with marginals gamma distributions. Particularly, the bivariate density considered for this example is given by

$$f(x, y \mid \omega_1, \omega_2, \omega_C) = c_t(F_X(x \mid \omega_1), F_Y(y \mid \omega_2) \mid \omega_C) \times Ga(x \mid \omega_1)\, Ga(y \mid \omega_2), \tag{25}$$

where $c_t(\cdot, \cdot|\cdot)$ is the copula density defined in (9) for $m = 2$ with $\omega_C = (\nu, \rho)$.

We simulated a sample $D_n = \{(x_1, y_1), \dots (x_n, y_n)\}$ of size $n = 500$ from the model (25) where $\omega_1 = (5, 4)$, $\omega_2 = (2, 1.5)$ and $\omega_C = (3, 0.7)$. Figure 3 shows the simulated data.

For this example, we use the next prior specification:

$$\pi(\omega_1, \omega_2, \omega_C) = Ga(\nu - 2 \mid a_3, b_3)\, U(\rho \mid -1, 1) \prod_{j=1}^{2} Ga(\alpha_j \mid a_j, b_j)\, Ga(\beta_j \mid c_j, d_j),$$

where $U(\cdot \mid -1, 1)$ is the uniform density on interval (-1,1), $a_j = 1, b_j = 0.001$ for $j = 1, 2, 3$, and $c_j = 1, d_j = 0.001$ for $j = 1, 2$.

We use the conditional approach described in previous section for sampling from the corresponding posterior distribution $\pi(\omega_1, \omega_2, \omega_C \mid D_n)$. For the above, we use a similar adaptative rejection algorithm within a Gibbs sampling as the Example 1.

Figure 4 shows the component-wise marginal distributions for the vectors ω_1, ω_2 and ω_C. On the other hand, the third column of Table 2 shows the 95% posterior credible intervals for each elements of the vectors ω_1, ω_2 and ω_C, using the

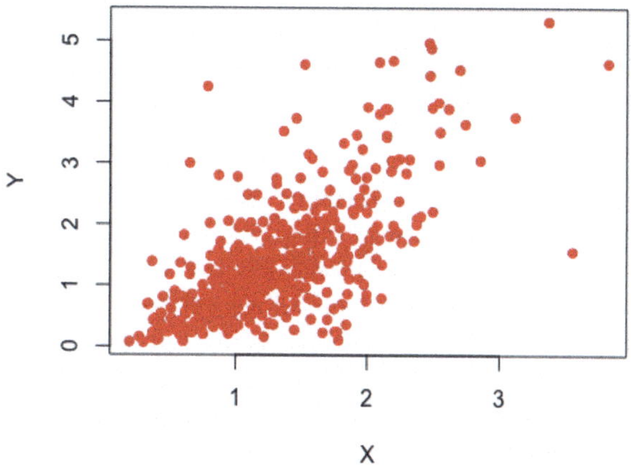

Fig. 3 Sample $\{(x_i, y_i)\}_{1 \le i \le n}$ of size $n = 500$ from model $f(x, y \mid \omega_1, \omega_2, \omega_c)$, where $\omega_1 = (5, 4)$, $\omega_2 = (2, 1.5)$ and $\omega_C = (3, 0.7)$

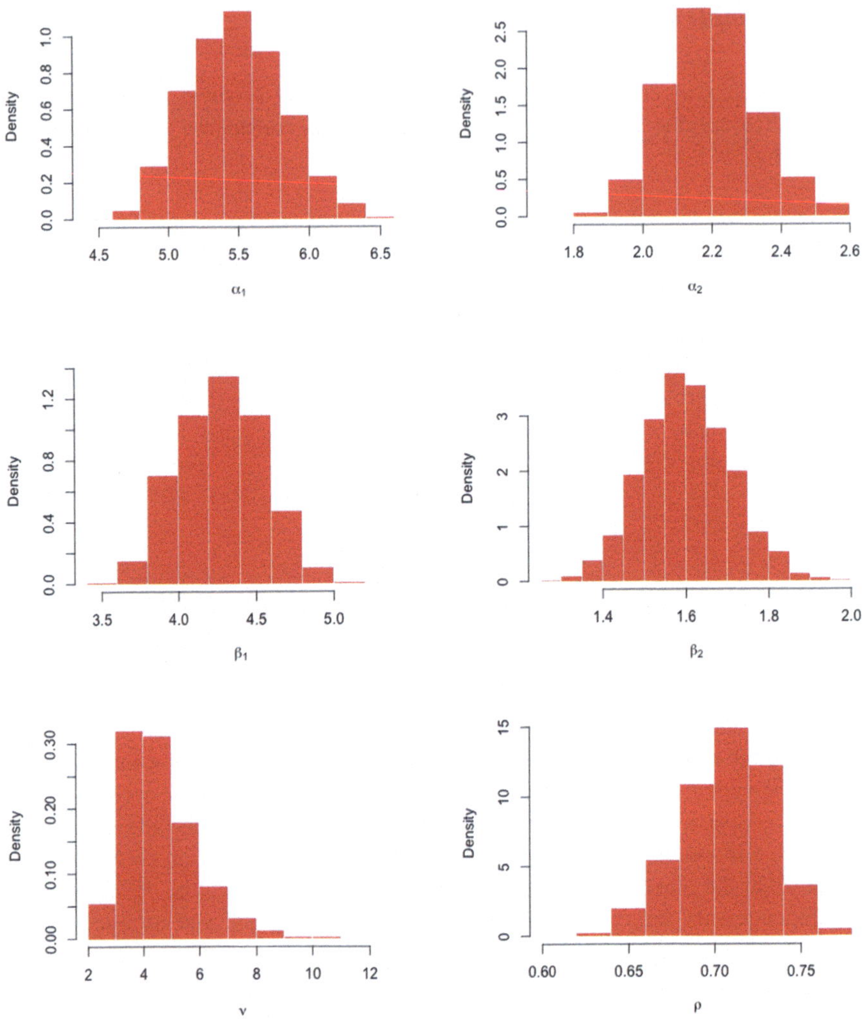

Fig. 4 Posterior distributions for component-wise of parameter vectors ω_1, ω_2 and ω_C for Example 2

conditional approach. It can be seen that the true value of each of the parameters is well inside the corresponding credible interval.

As a way to contrast the inferences obtained under the conditional approach, we implemented two additional algorithms. Columns 4 and 5 of Table 2 show the respective credibility intervals obtained under a Gibbs sampling and an algorithm t-walk (see [3]). It can be observed that inferences produced by the three algorithms are comparable. However, the conditional approach is more efficient in terms of estimation times and therefore a lower computational cost.

Table 2 95% credible intervals for each parameters of vectors ω_1, ω_2, and ω_C of the model of Example 2

Parameter	Real value	Credibility intervals		
		Condicional (255 sec)	Gibbs (1631 sec)	t-walk (1988 sec)
α_1	5	(4.19, 5.32)	(4.74, 5.97)	(4.42, 5.57)
α_2	2	(1.76, 2.22)	(1.97, 2.45)	(1.80, 2.27)
β_1	4	(3.24, 4.16)	(3.71, 4.71)	(3.53, 4.55)
β_2	1.5	(1.18, 1.54)	(1.42, 1.81)	(1.37, 1.78)
ρ	0.7	(0.66, 0.76)	(0.60, 0.72)	(0.60, 0.72)
ν	3	(2.00, 3.96)	(2.33, 6.72)	(2.00, 3.32)

Based on all the results obtained earlier, it can be concluded that under this conditional approach, appropriate Bayesian inferences can be obtained for these types of models defined by a copula structure.

5 Concluding Remarks

In this work, we show an advantageous way to carry out Bayesian inferences in models defined through copulas. This methodology is based on the conditional method proposed by Rosenblatt [11] to sample multivariate probability distributions. Although there are different ways to carry out Bayesian inferences in copula models, the *conditional approach* proposed in this work takes advantage of the structure of a copula model. That means, that approach offers a more efficient sampling of the posterior distribution of model parameters. The approach presented in this work turns out to be efficient; however, to the best of our knowledge, recent work on inference for copula models does not use this approach naturally. Finally, it should be mentioned that this conditional approach can be used in a direct way in models with larger dimensions and/or with more general marginal distributions such as those derived from non-parametric context. Work is currently in progress on these problems.

Acknowledgments The authors were partially supported from CONAHCYT-Mexico, the third through Sistema Nacional de Investigadores and the others by Sistema Nacional de Becas de Posgrado. Support from the Department of Mathematics of the Metropolitan Autonomous University, Iztapalapa Unit is also gratefully acknowledged. Finally, the authors wish to thank two anonymous reviewers for their valuable comments and suggestions.

Appendix

***Proof of Theorem** 1.* Let $x_1 \in \mathbb{R}^l$ be a particular value of the variable X_1. Then

$$\begin{aligned} f(X_2|X_1 = x_1) &= \frac{f(x_1, X_2)}{\int f(x_1, X_2)\, dX_2} \\ &\propto \frac{g_1(x_1, X_2) \times g_2(x_1)}{g_2(x_1) \times \int g_1(x_1, X_2)\, dX_2} \\ &= \frac{g_1(x_1, X_2)}{\int g_1(x_1, X_2)\, dX_2} \\ &\propto g_1(x_1, X_2). \end{aligned}$$

The aforementioned shows the first statement of the theorem. To prove the second statement, let us note that

$$\begin{aligned} f(x_1, X_2) &= f(X_2|X_1 = x_1) \times f(x_1) \\ &\propto g_1(x_1, X_2) \times f(x_1). \end{aligned}$$

Additionally, since by hypothesis $f(X_1, X_2) \propto g_1(X_1, X_2) \times g_2(X_1)$, then you have to

$$g_1(x_1, X_2) \times f(x_1) \propto g_1(x_1, X_2) \times g_2(x_1),$$

which implies that $f(x_1) \propto g_2(x_1)$. Since the above occurs for arbitrary x_1, then it is concluded that

$$f(X_1) \propto g_2(X_1),$$

which proves the second statement.

References

1. Ausin, M. C. and Lopes, H. F. (2010). Time-varying joint distribution through copulas. *Computational Statistics and Data Analysis*, 54, 2383–2399.
2. Biswabrata, P. and Debasis, K. (2011). Bayes estimation and prediction of the two-parameter gamma distribution. *Journal of Statistical Computation and Simulation*, 81:9, 1187–1198.
3. Christen, J. A. and Fox, C. (2010). A general purpose sampling algorithm for continuous distributions (the t-walk).
4. Demarta, S. and McNeil, A. (2005). The *t* copula and related copulas. *International Statistical Review*, 73, 1,111–129.
5. Gilks, W. R. and Wild, P. (1992). Adaptive rejection sampling for Gibbs sampling. *Applied Statistics*, 41, 337–348.
6. Gilks, W. R., Best, N. G. and Tan, K. K. C. (1995). Adaptive rejection Metropolis sampling. *Applied Statistics*, 44, 455–472.
7. Hoeffding, W. (1940). Massstabinvariante Korrelationstheorie. Schriften des Mathematischen Seminars und des Instituts für Angewandte Mathematik der Universität Berlin, 5, 181–233.
8. Johnson, M. E. (1987). *Multivariate Statistical Simulation*, John Wiley and Sons, Inc.

9. Nelsen, R. (2006). *An Introduction to Copulas*, Second Edition, Springer.
10. Pitt, M., Chan, D. and Kohn, R. (2006). Efficient Bayesian inference for Gaussian copula regression models. *Biometrika*, 93(3), 537–554.
11. Rosenblatt, M. (1952). Remarks on a multivariate transformation, *The Annals of Mathematical Statistics*, Vol. 23, No. 3 (Sep., 1952), pp. 470–472.
12. Silva, R. and Lopes, H. (2008). Copula, marginal distributions and model selection: a Bayesian note. *Statistics and Computing*, 18, 313–320.
13. Sklar, A. (1959). Fonctions de répartition à n dimensions et leurs marges. Publications de l'Institut de Statistique de l'Université de Paris, 8, 229–231.
14. Smith, M. S. (2011). Bayesian approaches to copula modelling. *arXiv preprint arXiv*:1112.4204.
15. Smith, M. S. and Klein, N. (2020). Bayesian Inference for Regression Copulas. *Journal of Business & Economic Statistics*, Vol 39, No. 3, 712–728.

A New Bivariate Model Based on Gamma Distributions

Gabriel Núñez-Antonio and Joel Montesinos-Vázquez

Abstract Recently, analysis of *circular data* has become a very dynamic development area. This goes from the construction of new probability models that describe this type of data, to the proposal of methods to carry out both frequentist and Bayesian inferences in the proposed models. One approach that has proved attractive for its simplicity in constructing circular distributions is the radial projection method. That consists of radially projecting a bivariate density, defined in $\mathbb{R}^2$, onto the unit circle $\mathbb{S}$. Although that method of construction is relatively simple, the crucial point is to have families of distributions initially defined in $\mathbb{R}^2$ that allow generating sufficiently flexible circular densities to describe data on $\mathbb{S}$. An additional problem arises when we want to have distributions defined only on a subset of $\mathbb{S}$. This chapter presents a new bivariate model defined in $\mathbb{R}^2$, as well as the corresponding circular model obtained from its radial projection. It should be mentioned that the projected model obtained is able to describe circular data defined only in the first orthant of $\mathbb{S}$ and is more flexible than those proposed in the literature. It also shows how to carry out Bayesian inferences for all parameters of the proposed bivariate model and describes some properties of the obtained circular model.

Keywords Radial projection · Gibbs sampler · circular data.

1 Introduction

In literature, there are variables or observations that naturally represent directions. These observations are common in biology, earth sciences, meteorology, and envi-

---

G. Núñez-Antonio
Departamento de Matemáticas, Universidad Autónoma Metropolitana - Unidad Iztapalapa, Mexico City, México

J. Montesinos-Vázquez (✉)
Universidad Autónoma Metropolitana, Unidad Iztapalapa, Mexico City, México

J. A. Christen et al. (eds.), *Statistics, Society and Environment*, Springer Proceedings in Mathematics & Statistics 479, https://doi.org/10.1007/978-3-031-78401-9_7

ronmental sciences. The directional data are observations that could be represented as unitary vectors in the p−dimensional euclidean space, so that they also could be described by $p - 1$ angles. Particularly, when $p = 2$ and $p = 3$ are referred as circular data and spherical data, respectively. For a survey from the frequentist perspective, the reader is referred to [3, 6, 8, 9, 12]. Arnold y SenGupta [1] shows a general frame of applications in ecology and environmental sciences. On the other hand, [7] presents a review of methods used in recent decades to deal with circular data.

In recent years, the development of statistical methods to analyze this type of directional data has had a new impetus. Those new contributions range from appropriate graphical methods to the proposal of new statistical models for the description of this type of observations with a directional nature. The reader can review, for example, the works of [2, 10, 11, 13] and the references therein included.

Although there are several ways to define probability distributions for circular variables, one of the simplest is to project radially a probability density originally defined in $\mathbb{R}^2$. An important case is the projected normal distribution, which is generated by the projection of a bivariate vector, which follows a normal probability distribution. It is important to mention that even when in recent years the normal projected model has received special attention, in general the study of models generated by radial projection has been overlooked. On the other hand, the circular models for variables restricted to subsets of $\mathbb{S}$ have been disregarded. For example, the *axial data*, which are represented as angular data on $(0, \pi]$. Although such observations are recorded as angles, circular probability models defined on a complete unit circle are inappropriate to describe them [1]. In addition, examples of data defined as angular observations on $[0, \frac{\pi}{2}]$ are mentioned in [10]. For these bounded observations, few probability models have been developed to date.

In this work, we introduce a probability model in $\mathbb{R}^2$, which when projected radially produces densities over the first quadrant of the unit circle. That model is based on a bivariate density defined from univariate Gamma densities with a non-independence structure. We derive all complete conditional densities necessary to perform Bayesian inferences for all parameters of the proposed model. Finally, the performance of the proposed methodology is shown through two simulated examples and a real example.

2 A New Family of Bivariate Densities

A versatile probability distribution defined on the positive reals is the gamma distribution. Thus, a proposal to define probability distributions on the first quadrant of $\mathbb{R}^2$ is to consider a joint density defined as

$$f(y_1, y_2|\boldsymbol{\nu}) = Gamma(y_1|\alpha_1, \beta_1) \times Gamma(y_2|\alpha_2, \beta_2) \tag{1}$$

where $\boldsymbol{\nu} = (\alpha_1, \alpha_2, \beta_1, \beta_2)$. The model (1) is precisely the model defined by Núñez-Antonio and Geneyro [10]. It should be noted that although that model produces a projected model, which is flexible to describe circular data defined in the first quadrant of $\mathbb{S}$, there are patterns or behaviors that it cannot yet capture properly. That is due to the structure of independence inherent in the model. Therefore, a generalization is to consider the next bivariate model

$$f(y_1, y_2|\boldsymbol{\eta}) \propto Gamma(y_1|\alpha_1, \beta_1) \times Gamma(y_2|\alpha_2, \beta_2) \times (c_1 y_1^{m_1} + c_2 y_2^{m_2}), \qquad (2)$$

where $\boldsymbol{\eta} = (\alpha_1, \beta_1, \alpha_2, \beta_2, c_1, c_2, m_1, m_2)$; and $m_1, m_2, c_1, c_2 \in \mathbb{R}^+$.

Although the model (2) no longer has an independent structure for the vector $\boldsymbol{Y} = (Y_1, Y_2)$, it should be noted that the parameters c_1 and c_2 do not produce a significant gain in the description of the behaviors of the circular model it generates. Therefore, to construct a more flexible circular model defined only in the first quadrant of the unit circle, in this chapter, we propose the following bivariate model based on gamma densities.

$$f(y_1, y_2|\boldsymbol{\gamma}) \propto Gamma(y_1|\alpha_1, \beta_1) \times Gamma(y_2|\alpha_2, \beta_2) \times (y_1^{m_1} + y_2^{m_2}), \qquad (3)$$

where $\boldsymbol{\gamma} = (\alpha_1, \beta_1, \alpha_2, \beta_2, m_1, m_2)$; $\alpha_1, \beta_1, \alpha_2, \beta_2, m_1, m_2 \in \mathbb{R}^+$.

If we consider the transformation to polar coordinates $(Y_1, Y_2) \rightarrow (R, \Theta)$, where $R = ||\boldsymbol{Y}||$, then we can derive, through the variable change theorem, the joint density of R, and Θ, $f_{R,\Theta}(r, \theta)$, as follows.

$$f_{R,\Theta}(r, \theta\ |\boldsymbol{\gamma}) = f_{Y_1,Y_2}((r, \theta)|\boldsymbol{\gamma}) \times |\boldsymbol{J}(r, \theta)|,$$

where $r = |\boldsymbol{J}(r, \theta)|$ is the Jacobian of the inverse transformation. Thus, it turns out

$$\begin{aligned} f(r, \theta|\boldsymbol{\gamma}) = \kappa\ & Gamma(r\cos\theta|\alpha_1, \beta_1) \times Gamma(r\sin\theta|\alpha_2, \beta_2) \\ & \times((r\cos\theta)^{m_1} + (r\sin\theta)^{m_2}) \times r, \end{aligned}$$

where

$$\kappa = \left[\frac{\Gamma(\alpha_1 + m_1)}{\Gamma(\alpha_1)\beta_1^{m_1}} + \frac{\Gamma(\alpha_2 + m_2)}{\Gamma(\alpha_2)\beta_2^{m_2}}\right]^{-1}.$$

From the joint density $f(r, \theta|\boldsymbol{\gamma})$, we can get the density $h(\theta|\boldsymbol{\gamma})$, which is a density for the circular variable Θ defined only in the first quadrant of the unit circle $\mathbb{S}$. That density $h(\theta|\boldsymbol{\gamma})$, obtained under radial projection, turns out to be

$$\begin{aligned} h(\theta|\boldsymbol{\gamma}) &= \kappa\left[\frac{\Gamma(m_1 + \alpha_1)}{\Gamma(\alpha_1)\beta_1^{m_1}} GP(\theta|\alpha_1 + m_1, \alpha_2, \beta_1, \beta_2)\right] \\ &\quad + \kappa\left[\frac{\Gamma(m_2 + \alpha_2)}{\Gamma(\alpha_2)\beta_2^{m_2}} GP(\theta|\alpha_1, \alpha_2 + m_2, \beta_1, \beta_2)\right], \end{aligned} \qquad (4)$$

where

$$GP(\theta|a_1,a_2,\beta_1,\beta_2) = \frac{\beta_1^{a_1}\beta_2^{a_2}\cos\theta^{a_1-1}\sin\theta^{a_2-1}}{B(a_1,a_2)[\beta_1\cos\theta+\beta_2\sin\theta]^{a_1+a_2}}$$

is the *projected Gamma distribution* defined by Núñez-Antonio and Geneyro [10].

3 The Proposed Model

In this work, as a model on the first quadrant of $\mathbb{R}^2$, we propose the model

$$f(y_1,y_2|\boldsymbol{\gamma}) \propto Gamma(y_1|\alpha_1,\beta_1) \times Gamma(y_2|\alpha_2,\beta_2) \times (y_1^{m_1}+y_2^{m_2}).$$

Particularly, that joint density induces a model for the circular variable θ given by (4).

Figure 1 shows some Gamma protected densities (see [10]). As aforementioned, the model $h(\theta|\boldsymbol{\gamma})$ turns out to be more flexible than that proposed by Núñez-Antonio y Geneyro [10]. That means, there are patterns that $h(\theta|\boldsymbol{\gamma})$ describes better than the model $GP(\theta|a_1,a_2,\beta_1,\beta_2)$. For example, $GP(\theta|\cdot)$ is able to describe the bimodality represented in Fig. 1b, but it is not able to describe the bimodal behavior of Fig. 2, as model $h(\theta|\cdot)$ does. As another example, the bimodality that model $h(\theta|\cdot)$ describes in Fig. 4 is not possible to describe with $GP(\theta|\cdot)$. Thus, the model $h(\theta|\cdot)$ may be a better option to describe circular data when more pronounced *valleys* are present in the data to be analyzed.

In a practical context, the use of the circular model $h(\theta|\boldsymbol{\gamma})$ depends greatly on being able to make inferences for the bivariate model $f(y_1,y_2|\boldsymbol{\gamma})$. In particular, the ability to make inferences for all parameters of the vector $\boldsymbol{\gamma} = (\alpha_1,\beta_1,\alpha_2,\beta_2,m_1,m_2)$. In the next section, we show how to carry out these inferences.

4 Bayesian Inference

There are several approaches to making inferences about the parameters of a statistical model. In this work, we use a Bayesian approach and Markov Chain Monte Carlo (MCMC) methods to sample from final distribution of all parameters involved. Next, we present all the necessary elements.

If $\boldsymbol{D_n} = \{(y_{1i},y_{2i})\}_{i=1}^n$ is a sample of size n of the model (3), then the likelihood turns out to be

$$L(\boldsymbol{D_n}|\boldsymbol{\gamma}) = \kappa^n \prod_{i=1}^n Gamma(y_{1i}|\alpha_1,\beta_1)Gamma(y_{2i}|\alpha_2,\beta_2)(y_{1i}^{m_1}+y_{2i}^{m_2}).$$

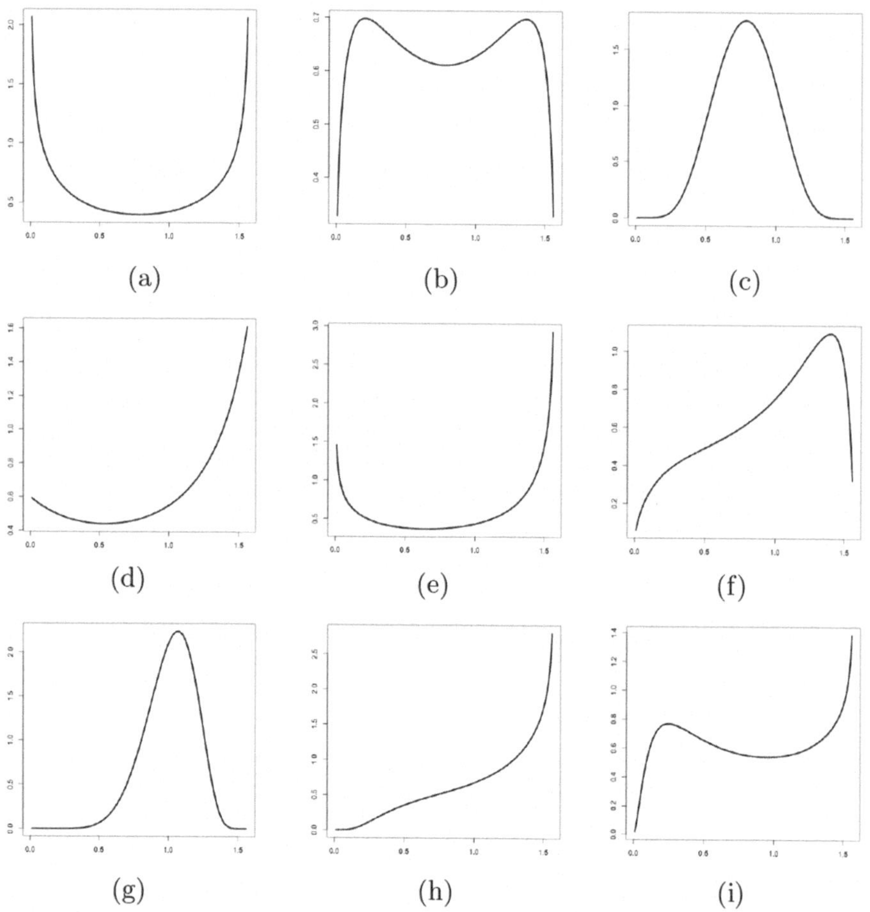

Fig. 1 Possible patterns produced by the projected Gamma model, $GP(\theta|\cdot)$, proposed by Núñez-Antonio y Geneyro [10]. (**a**) $\alpha_1 = \alpha_2 = 1, \beta_2 = 1$. (**b**) $\alpha_1 = \alpha_2 = 1.4, \beta_2 = 1$. (**c**) $\alpha_1 = \alpha_2 = 10, \beta_2 = 1$. (**d**) $\alpha_1 = \alpha_2 = 1, \beta_2 = 0.6$. (**e**) $\alpha_1 = \alpha_2 = 0.7, \beta_2 = 0.6$. (**f**) $\alpha_1 = \alpha_2 = 1.5, \beta_2 = 0.6$. (**g**) $\alpha_1 = \alpha_2 = 12, \beta_2 = 0.6$. (**h**) $\alpha_1 = 0.8, \alpha_2 = 10, \beta_2 = 7$. (**i**) $\alpha_1 = 0.8, \alpha_2 = 3, \beta_2 = 5$

We propose the following prior specification for the parameter vector $\boldsymbol{\gamma}$

$$\pi(\boldsymbol{\gamma}) = \pi_1(\alpha_1)\pi_2(\alpha_2)\pi_3(\beta_1)\pi_4(\beta_2)\pi_5(m_1)\pi_6(m_2),$$

where

$$\pi_1(\alpha_1) = Gamma(a_1, b_1),\ \ \pi_2(\alpha_2) = Gamma(a_2, b_2),$$

$$\pi_3(\beta_1) = Gamma(c_1, d_1),\ \ \pi_4(\beta_2) = Gamma(c_2, d_2),$$

$$\pi_5(m_1) = Gamma(e_1, f_1), \ \pi_6(m_2) = Gamma(e_2, f_2).$$

Therefore, the joint final distribution turns out to be

$$f(\boldsymbol{\gamma}|\boldsymbol{D_n}) = \kappa^n \prod_{i=1}^{n} Gamma(y_{1i}|\alpha_1, \beta_1) Gamma(y_{2i}|\alpha_2, \beta_2)(y_{1i}^{m_1} + y_{2i}^{m_2})$$
$$\times \prod_{j=1}^{2} Gamma(\alpha_j|a_j, b_j) Gamma(\beta_j|c_j, d_j) Gamma(m_j|e_j, f_j). \tag{5}$$

Since that joint final distribution is difficult to analyze analytically or numerically, to make inferences about the parameters of the model, we propose to implement a Gibbs sampling ([4]). For the aforementioned, it needs to specify all complete conditional densities from (5). Subsequently, we must be able to sample from all of those complete conditionals. Next, all the full conditional distributions needed to implement a Gibbs sampling are derived.

4.1 Posterior Conditional Densities

The full conditional density for the parameter α_1 is given by

$$f(\alpha_1|\cdot) = \frac{f(\boldsymbol{\gamma}|\boldsymbol{D_n})}{\int_0^\infty f(\boldsymbol{\gamma}|\boldsymbol{D_n})\, d\alpha_1}.$$

After some calculations, we get that

$$f(\alpha_1|\cdot) \propto \kappa^n \prod_{i=1}^{n} Gamma(y_{1i}|\alpha_1, \beta_1) \times \pi_1(\alpha_1). \tag{6}$$

Similarly, the full conditional densities for the parameters α_2, β_1, β_2, m_1 and m_2 are obtained. These posterior conditionals are given up to a constant of proportionality by

$$f(\alpha_2|\cdot) \propto \kappa^n \prod_{i=1}^{n} Gamma(y_{2i}|\alpha_2, \beta_2) \times \pi_2(\alpha_2), \tag{7}$$

$$f(\beta_1|\cdot) \propto \kappa^n \prod_{i=1}^{n} Gamma(y_{1i}|\alpha_1, \beta_1) \times \pi_3(\beta_1), \tag{8}$$

	Gibbs sampling
Step 1.	Assign initial values for the vector $\boldsymbol{\gamma}$
Step 2.	Sampling α_1 from $f(\alpha_1\|\cdot)$
Step 3.	Sampling α_2 from $f(\alpha_2\|\cdot)$
Step 4.	Sampling β_1 from $f(\beta_1\|\cdot)$
Step 5.	Sampling β_2 from $f(\beta_2\|\cdot)$
Step 6.	Sampling m_1 from $f(m_1\|\cdot)$
Step 7.	Sampling m_2 from $f(m_2\|\cdot)$
Step 8.	Repeat until convergence is achieved.

$$f(\beta_2|\cdot) \propto \kappa^n \prod_{i=1}^{n} Gamma(y_{2i}|\alpha_2, \beta_2) \times \pi_4(\beta_2), \tag{9}$$

$$f(m_1|\cdot) \propto \kappa^n \prod_{i=1}^{n} (y_{1i}^{m_1} + y_{2i}^{m_2}) \times \pi_5(m_1), \tag{10}$$

$$f(m_2|\cdot) \propto \kappa^n \times \prod_{i=1}^{n} (y_{1i}^{m_1} + y_{2i}^{m_2}) \times \pi_6(m_2). \tag{11}$$

Once all posterior conditional densities (6)–(11) have been obtained, it is possible to implement a Gibbs sampling. The corresponding algorithm is shown into the Gibbs sampling frame.

It should be mentioned that to sample the densities in each of steps 2–5, an *Adaptative Rejection Metropolis Sampling* (ARMS) algorithm was implemented. We use the arms function from the armspp R-library. For further reference of that algorithm, the reader may refer to [5]. Additionally, for steps 6 and 7, we use a Metropolis–Hastings algorithm (see, for example [14], and references included therein).

5 Simulation Studies

This section presents three examples showing the Bayesian inferences obtained for the proposed model $f(y_1, y_2|\boldsymbol{\gamma})$.

5.1 Example 1

In this example, we simulate a sample of size $n = 500$ from the model $f(y_1, y_2|\boldsymbol{\gamma})$, where $\alpha_1 = 2.5$, $\alpha_2 = 2.5$, $\beta_1 = 7$, $\beta_2 = 7$, $m_1 = 7$ and $m_2 = 7.05$. The

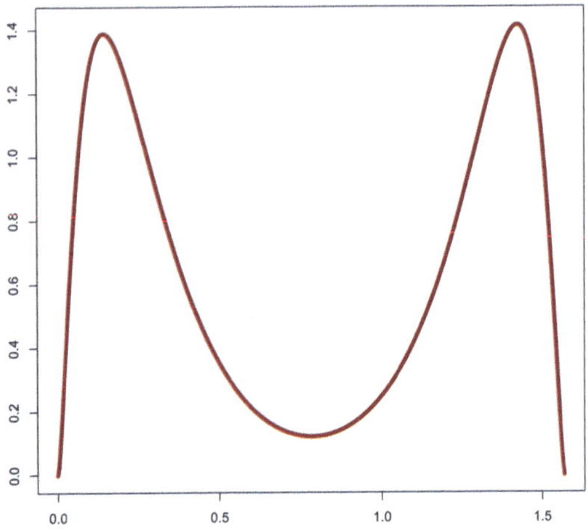

Fig. 2 The corresponding density of the circular variable Θ associated with the cartesian model $f(y_1, y_2|\boldsymbol{\gamma})$ in Example 1

aforementioned specification produces a bimodal density for the circular variable Θ, which is defined over the first quadrant of the unit circle $\mathbb{S}$. Figure 2 shows a graph of that density.

We used the next specification

$$\begin{aligned}\pi(\boldsymbol{\gamma}) = & \; Gamma(\alpha_1|a_1 = 1, b_1 = \tfrac{1}{3}) \times Gamma(\alpha_2|a_2 = 1, b_2 = \tfrac{1}{3}) \\ & \times Gamma(\beta_1|c_1 = 1, d_1 = \tfrac{1}{3}) \times Gamma(\beta_2|c_2 = 1, d_2 = \tfrac{1}{3}) \\ & \times Gamma(m_1|e_1 = 1, f_1 = \tfrac{1}{3}) \times Gamma(m_2|e_2 = 1, f_2 = \tfrac{1}{3}),\end{aligned}$$

as prior distribution. That prior specification produces distributions with a large variance and thus turns out to be non-informative. We apply the simulation algorithm described in the previous section to get a sample of size 1,000 from the posterior distribution $f(\boldsymbol{\gamma}|\boldsymbol{D_n})$. Convergence diagnostics led us to stop the process after 60,000 iterations, discarding the first 30,000 as burn-in. From the remaining 30,000 iterations, we kept one observation every 30 iterations as part of the final sample. It should be mentioned that the time required for the MCMC run described here (using a standard 64-bit computer with a 2.8 GHz CPU and 128 GB of RAM) was 18.5 min. The component-wise posterior marginal distributions for the vector $\boldsymbol{\gamma}$ are shown in Fig. 3. In addition, Table 1 shows the 95% posterior credible intervals for all parameters involved in the model. It can be noted that the true value of each of the parameters is well inside the corresponding credible interval. Moreover, given the shape of the final distribution of each of the parameters (Fig. 3), we can take as point estimates of parameters the mode of the corresponding posterior distributions, which are given by $\hat{\alpha_1} = 2.47$, $\hat{\alpha_2} = 2.47$, $\hat{\beta}_1 = 7.01$, $\hat{\beta}_2 = 6.95$, $\hat{m}_1 = 6.99$, and $\hat{m}_2 = 7.28$.

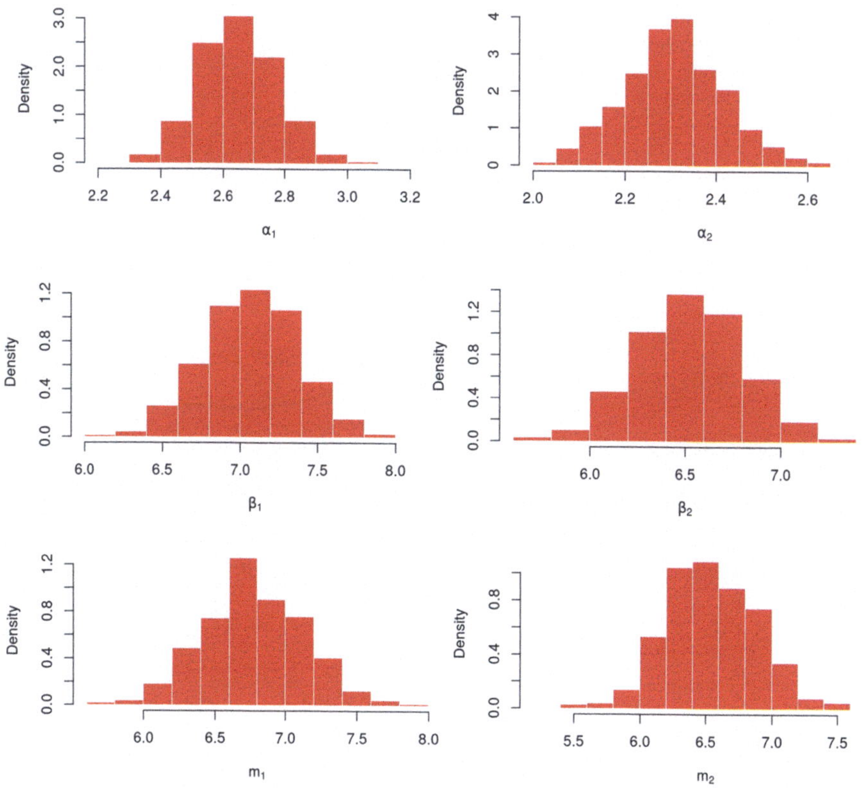

Fig. 3 Posterior distributions for component-wise of the parameter vector $\boldsymbol{\gamma}$

Table 1 95% credible intervals for parameters of the model in Example 1

Parameter	Actual value	Credible interval
α_1	2.5	(2.41, 2.90)
α_2	2.5	(2.08, 2.51)
β_1	7	(6.45, 7.63)
β_2	7	(5.95, 7.03)
m_1	7	(6.13, 7.48)
m_2	7.05	(5.92, 7.23)

5.2 *Example 2*

For this second example, we simulate a sample of size $n = 500$ from model $f(y_1, y_2|\boldsymbol{\gamma})$, where $\alpha_1 = 1.5$, $\alpha_2 = 1.5$, $\beta_1 = 4$, $\beta_2 = 5$, $m_1 = 7$ and $m_2 = 8$. In this case, the aforementioned specification produces a bimodal and asymmetric density for the corresponding circular variable Θ. Figure 4 shows a graph of that density. In this example, we use the same non-informative prior specificaction as in Example 1. A sample of size 1,000 of the posterior distribution was obtained

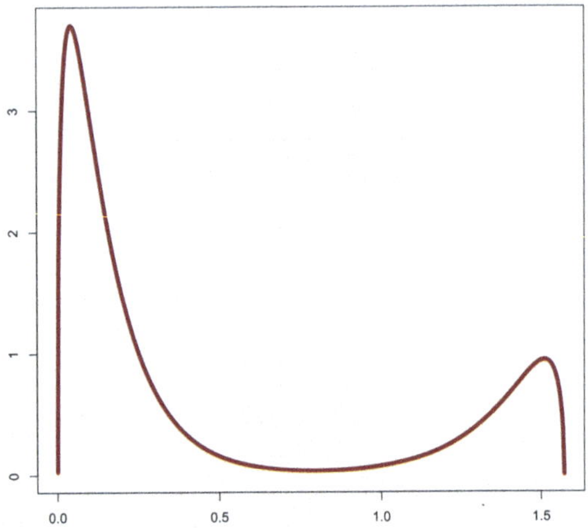

Fig. 4 Density for the circular variable Θ generated from cartesian model $f(y_1, y_2|\boldsymbol{\gamma})$ in Example 2

Table 2 95% credible intervals for parameters of the model in Example 2

Parameter	Actual value	Credible interval $b_j = d_j = f_j = \frac{1}{3}$	Credible interval $b_j = d_j = f_j = 0.01$	Credible interval $b_j = d_j = f_j = 0.001$
α_1	1.5	(1.25, 1.62)	(1.27, 1.64)	(1.28, 1.63)
α_2	1.5	(1.42, 1.72)	(1.44, 1.76)	(1.41, 1.72)
β_1	4	(3.31, 4.24)	(3.35, 4.29)	(3.36, 4.28)
β_2	5	(4.64, 5.80)	(4.60, 5.83)	(4.67, 5.85)
m_1	7	(5.52, 7.34)	(5.56, 7.44)	(5.56, 7.32)
m_2	8	(6.98, 9.14)	(7.12, 9.44)	(7.07, 9.35)

through the corresponding Gibbs sampler. For this example, a burn-in period of 40,000 iterations was established, and a lag of 30 iterations between each element to final sample was considered. The computational process took 20 min. Column 3 of Table 2 shows the 95% corresponding credible intervals for each parameters involved in the model. It can be noted that the true value of each of the model parameters is well inside the corresponding credible interval.

As a way to evaluate the sensitivity of the obtained inferences, with respect to the specification of the initial distributions used, we apply the proposed methodology under other prior specifications non-informative. Columns 4 and 5 of Table 2 show the corresponding credibility intervals. In particular, it can be observed that the inferences obtained are not sensitive to the specification of the non-informative prior distributions.

5.3 Example 3

In this example, we used data from 88 patients with primary brain tumors, see [15]. Two variables were analyzed: gross tumor volume (*gtv*), measured in cm^3, and survival time (*months*), measured in months. Figure 5 shows a scatter plot of data associated with these two variables. As a complementary analysis to that presented in [15], we implement the methodology proposed in this work to describe the variables *gtv* and *months*. In particular, we produce a sample of observations of the corresponding joint distribution for the variables *gtv* and *months*.

In this example, we use a prior specification with hyperparameters $a_i = c_i = e_i = 1$ and $b_i = d_i = f_i = 0.001$. A sample of size 1,000 of the posterior distribution was obtained through the corresponding Gibbs sampler. For this example, a burn-in period of 30,000 iterations was established, and a lag of 30 iterations between each element to final sample was considered. The computational process took 5 min.

Figure 6 shows the corresponding marginal distributions estimated for each of the variables *gtv* and *months*. On the other hand, Fig. 7 displays the sample of size 1,000

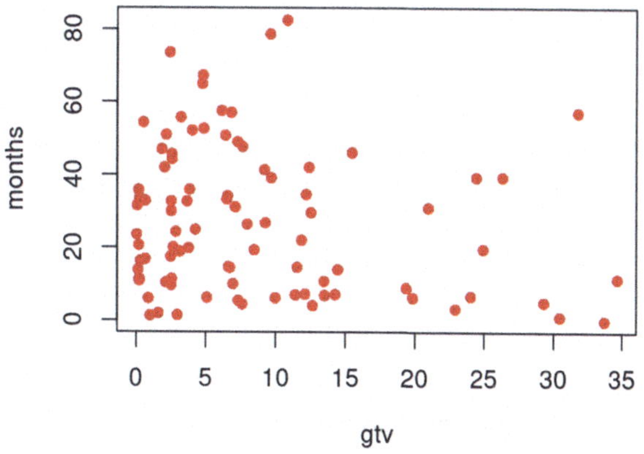

Fig. 5 Data from 88 patients with primary brain tumors: survival time (*moths*) versus gross tumor volume (*gtv*)

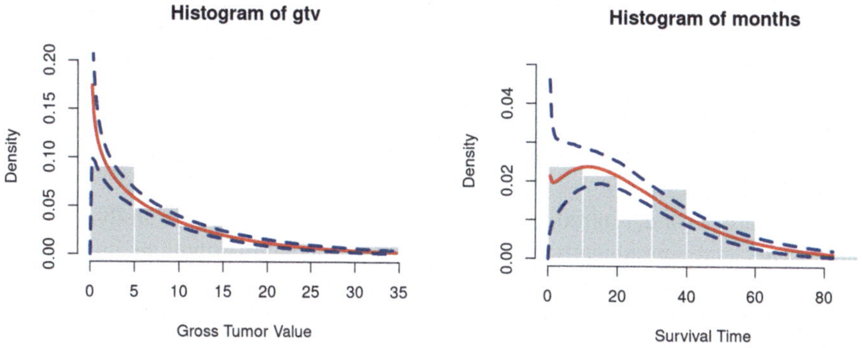

Fig. 6 Left: Marginal distribution of *gtv*. Right: Marginal distribution of *months*

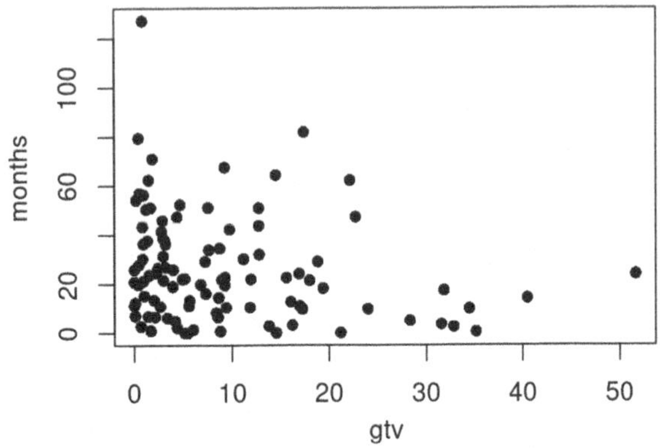

Fig. 7 Sample from the predictive distribution for Example 3

Table 3 95% credible intervals for parameters of the model in Example 3

Parameter	Credible interval
α_1	(0.54, 1.02)
α_2	(0.41, 1.43)
β_1	(0.08, 0.19)
β_2	(0.04, 0.09)
m_1	(0.31, 2.01)
m_2	(0.42, 2.02)

from the estimated joint distribution. In addition, Table 3 shows the corresponding 95% credible intervals for each parameters using the model (3).

All the analyses presented earlier suggest that the methodology proposed in this work is appropriate to make inferences for the proposed model $f(y_1, y_2|\boldsymbol{\gamma})$. This provides the basis for analyzing the projected model $h(\theta|\boldsymbol{\gamma})$ associated with the circular variable Θ, which is only defined in the first quadrant of the unit circle $\mathbb{S}$.

6 Concluding Remarks

In this work, we propose a new cartesian bivariate model whose definition is based on a non-independence structure of gamma distributions. This bivariate model, when projected radially, produces a circular distribution defined in the first quadrant over the unit circle. Additionally, this model is more flexible than the projected models proposed in the literature. We show how to perform Bayesian inferences using MCMC algorithms.

The bivariate model introduced in this work can be extended naturally to more dimensions, and consequently, it can be used to generate distributions on the unit sphere of dimension p. The aforementioned idea allows describing *directional* data restricted to subsets of the unit sphere. Work in these areas is being developed by the authors.

Acknowledgments The authors were partially supported from CONAHCYT-Mexico, the first through Sistema Nacional de Investigadores and the second by Sistema Nacional de Becas de Posgrado. Support from the Department of Mathematics of the Metropolitan Autonomous University, Iztapalapa Unit is also gratefully acknowledged. Finally, the authors are grateful to the anonymous reviewers for their detailed and insightful comments.

References

1. Arnold, B. C. and SenGupta, A. (2006). Recent advances in the analyses of directional data in ecological and environmental sciences. *Environmental and Ecological Statistics*, 13(3), 253–256.
2. D'Elia, A., Borgioli, C. and Scapini, F. (2001). Orientation of sandhoppers under natural conditions in repeated trials: an analysis using longitudinal directional data. *Estuarine, Coastal and Shelf Science*, 53(6), 839–847.
3. Fisher, N. I. (1995). *Statistical analysis of circular data.* Cambridge: University Press.
4. Geman, S. and Geman, D. (1984). Stochastic Relaxation, Gibbs Distributions, and the Bayesian Restoration of Images. *IEEE Transactions on Pattern Analysis and Machine Intelligence*, 6(6), 721–741.
5. Gilks, W. R., Best, N. G. and Tan, K. K. C. (1995). Adaptive rejection Metropolis sampling. *Applied Statistics*, 44, 455–472.
6. Jammalamadaka, S. R. and SenGupta, A. (2001). *Topics in circular statistics.* Singapore: World Scientific.
7. Lee, A. (2010). Circular data. *Wiley Interdisciplinary Reviews: Computational Statistics*, 2(4), 477–486.
8. Mardia, K. V. (1972). *Statistics of Directional Data.* London: Academic press.
9. Mardia, K. V. and Jupp, P. E. (2000). *Directional Statistics.* Chichester: Wiley.
10. Núñez-Antonio, G. and Geneyro, E. (2019). A multivariate projected Gamma model for directional data. *Communications in Statistics-Simulation and Computation*, 50(9), 2721–2742.
11. Paine, P. J., Preston, S. P., Tsagris, M. and Wood, A. T. (2018). An elliptically symmetric angular Gaussian distribution. *Statistics and Computing*, 28(3), 689–697.
12. Pewsey, A., Neuhäuser, M. and Ruxton, G. D. (2013). *Circular statistics in R.* Oxford: University Press.
13. Presnell, B., Morrison, S. P. and Littell, R. C. (1998). Projected multivariate linear models for directional data. *Journal of the American Statistical Association*, 93(443), 1068–1077.
14. Robert, C. P. (2016). The Metropolis-Hastings algorithm, *Wiley StatsRef: Statistics Reference Online*. John Wiley & Sons, Ltd.
15. Selingerova, I., Dolezelova, H., Horova, I., Katina, S. and Zelinka, J. (2016). Survival of patients with primary brain tumors: Comparison of two statistical approaches. *PLoS One*, 11(2): e0148733.

Mortality Estimation with Controlled Smoothing by Segments Applied to Potentially Insured Missing Persons in Mexico

Eliud Silva and Miguel Flores

Abstract We aim to estimate mortality in the insured Mexican market due to missing individuals who may have life insurance. To do that, we assess the potential impact of a future reform that could accelerate the process of obtaining Death Certificates of Presumption (CPDs) for life insurance beneficiaries. The worst scenarios of mortality have been generated, assuming that a certain percentage of missing individuals who have died were insured, based on data from 2016 to 2018. Our approach involves the implementation of controlled smoothing by segments, forecasting tools, and demographic calculations. The results suggest changes in mortality patterns, particularly at younger ages for both genders in collective insurance. Moreover, the results indicate that the observed mortality significantly exceeds the expected mortality for certain life stages, even without considering the potential impact of missing individuals who had life insurance. Consequently, the Mexican insurance sector should revise expected mortality and foresee additional consequences.

Keywords Life insurance · Missing persons · Life expectancy · Mexico

Author Eliud Silva, BSc Degree in Actuarial Sciences, Master's Degree in Demography, Doctorate in Mathematical Engineering, Actuarial Sciences Faculty, Universidad Anáhuac México

Author Miguel Flores, BSc Degree in Economics, Master's Degree in Economics, Doctorate in Demography, Tecnológico de Monterrey

E. Silva (✉)
Universidad Anáhuac México, Lomas Anáhuac, Mexico
e-mail: jose.silva@anahuac.mx

M. Flores
Departamento de Economía, Escuela de Ciencias Sociales y Gobierno, Tecnológico de Monterrey, Monterrey, Nuevo León, Mexico
e-mail: miguelflores@tec.mx

J. A. Christen et al. (eds.), *Statistics, Society and Environment*, Springer Proceedings in Mathematics & Statistics 479, https://doi.org/10.1007/978-3-031-78401-9_8

1 Introduction

According to the Comisión Nacional de Búsqueda [6], the number of missing persons in Mexico at the beginning of 2022 was 96,498 people. Specifically, 2019 and 2020 were two years with a large number of missing person cases, despite the almost generalized confinement imposed during 2020. Arista [2] argues that another pandemic has been occurring simultaneously, namely that of missing persons, suggesting that of the total number of missing and unaccounted for, many could coincide with the more than 30,000 unidentified bodies held at Mexico's forensic medical services. She adds that, in addition to the disappearances, there are clandestine graves. Likewise [16] points out that in many of the entities of the country, there is a limited capacity to recognize deceased bodies in the forensic services. In addition, the recognition process is particularly serious in several states, such as Mexico City, the State of Mexico, and Nuevo Leon.

The disappearance of potentially insured persons jeopardizes the ability to collect life insurance benefits. Some direct implications for the insured are: (a) their vital status is unknown, and their insurance may be cancelled for lack of payment; and (b) the beneficiaries do not have the documentation required to qualify for life insurance claims. Unfortunately, in the case of Mexico, there are no official figures on the number of potentially insured missing persons. This implies a challenge in terms of resource forecasting on the part of the insurance sector, especially if claims are to be made on a massive scale under the new regulations in this regard. On the other hand, there is evidence of a growing trend of homicides in the insured population based on annual reports of the Asociación Mexicana de Instituciones de Seguros [5]. Sadly, a recurring pattern is that when missing persons are located, they appear to be victims of homicides. Fuentes and Arellano [10] document the situation of violence prevailing in the country, which affects both the population with life insurance and the Mexican insurance sector itself.

The aim of this chapter is to estimate mortality assuming that all the missing insured have died for the case of Mexico, considering the years from 2016 to 2018, which is the period when information is available at the microdata level. For this purpose, we use the controlled smoothing by segments methodology proposed by Guerrero and Silva [12]. The estimates consider data from AMIS and the Secretaría de Gobernación (SEGOB). Based on the results obtained, it is considered relevant to reestimate the mortality rate currently used in Mexico, both for individual and collective life insurance, regulated by the Comisión Nacional de Seguros y Fianzas (CNSF). According to the calculations, there are notable mortality excesses in the various estimates that are made. Thus, the first section of the chapter deals with the problem of missing persons and provides some profiles and trends over time and in the different entities of the country based on the available information. Subsequently, the main ideas of the statistical method used are explained. Some scenarios are presented based on different assumptions, and the most important conclusions are outlined.

Some limitations are recognized for the realization of this work. On the one hand, we only have information at the microdata level during the period under study, i.e., for the missing persons recognized by the CNB, which is why we are studying cases from the 2016–2018 triennium and not more recent information. This database, with such a level of disaggregation, was no longer officially published as of 2018. Currently, even when the corresponding statistical information is available, it is at an aggregate level, which would require making demographic assumptions about the quotas and in turn could bias the estimates. On the other hand, AMIS also does not have a public census to indicate which policyholders are missing, at least considering their age, sex, and/or geographic location. This would provide greater certainty about the volume of the total number of missing policyholders, as well as the type of insurance they have.

2 Background

The insured who disappear undoubtedly cause impacts of different nature within their family. Without at least a death certificate, the claim is inadmissible. According to the current Mexican regulations, there are up to 5 years from the death of the insured to make the claim, if the event occurs during the term of the policy (Insurance and Bonding Companies Law, 2013). Among other requirements, a death certificate is the legal document necessary for the collection of life insurance in favor of the beneficiaries. In case of the disappearance of insured persons, the beneficiaries have to obtain the so-called Certificate of Presumed Death (CPD), which can take between 1.5 and 6 years, starting from the declaration of absence. Besides, the waiting time is not the same all over the country and varies according to local laws [21]. There are currently some legal loopholes with respect to how to claim life insurance for missing persons.

In the event of the disappearance of an insured, according to [20], there are few cases in which it has been possible to collect insurance payments. In order to succeed, it is necessary to file a complaint with the Public Prosecutor's Office (Ministerio Público) and thus obtain the so-called declaration of absence. To point out some deadlines, they range from 2 to 3 years of waiting time. If the process is successful, the relatives can request the CPD from the civil registry of their corresponding entity. In the United States, according to [8], there must be 7 years of he continuous time of disappearance without reasonable explanation, among other elements. However, if it is proven that the missing person is alive, the company is entitled to recover the sum insured plus interest. Dubey [9] stresses that, in other countries, in order for a missing person to be declared dead and for life insurance to be paid out, it is sufficient that they are recognized on a government official's list of presumed victims following a natural disaster.

The fact that there are insured missing persons, given the volume of cases in Mexico, is considered to be a latent risk situation for the insurance sector, at least in terms of solvency. In fact, the current mortality tables by sex, both for individual and collective life insurance, regulated by the CNSF, do not consider these surpluses. Of course, new tables would significantly affect the pricing of life insurance and its actuarial reserves. Furthermore, it is not implausible that a rule or law would suddenly be implemented that would oblige insurance companies to pay the respective insured amounts after a certain period of time has elapsed due to the disappearances. It should be noted that, in some states of the country, there are already initiatives in this direction. The following section provides some macro figures on missing persons and the insurance sector.

3 Material and Methods

3.1 Data

According to the [6], there are more than 96,000 missing persons nationwide. This figure in terms of volume is similar to the total population of some U.S. counties such as those shown in the following Table 1, according to the U.S. Census Bureau [22]. Likewise, that number of people is comparable to the respective registered population of some Mexican municipalities according to the final results of the Mexican Population Census for the year 2020 elaborated by INEGI [14]. It is a fact that in both the U.S. and Mexico, the populations of such geographic units have been altered due to demographic events derived from the COVID-19 pandemic during 2020 (see for example [15]).

As of May 2018, the country's official registry of missing persons was no longer published at the microdata level as it had been done until then; that is, in a publicly accessible database. This has implied that there is no information at the individual

Table 1 Population from selected counties and municipalities from the U.S. and México, 2020

City	State	Population
U.S.		
Ashtabula County	Ohio	97,574
Lafourche Parish	Los Ángeles	97,557
Bastrop County	Texas	97,216
Geauga County	Ohio	95,397
Cape May County	New Jersey	95,263
Rogers County	Oklahoma	95,240
Mexico		
Rio verde	San Luis Potosí	97,943
Cortázar	Guanajuato	97,928
San Mateo Atenco	Estado de México	97,418
Medellín de Bravo	Veracruz	95,202
Tamazunchale	San Luis Potosí	95,037
Salvatierra	Guanajuato	94,126

Source: Own elaboration with data from [22] and [14]

level of the victims from the past years to date. Some efforts have been made for its construction, as stated in [3], although it is emphasized that information of an official nature is aggregated. In turn, this situation means that, ultimately, information on the demographic and geographic profile of the new cases of missing persons is absent. It is likely that in the midst of the confinement of the COVID-19 pandemic, new patterns have emerged. It is for this reason that the present work focuses on the period closest to the information captured at the individual level, that is, from 2016 to 2018. Based on historical observations, and in order to have complete years, forecasts were made from May to December 2018.

As of April 2018, the existing official database has individual information on missing persons, both federal and common laws [18], which are used together to make some estimates in this document. The variables in the database of federal jurisdiction are: date, country, entity and municipality where the person was last seen, name(s), paternal surname, maternal surname, entity code, country of birth, nationality, sex, age, type of place, authority and entity, and date on which the report was received. In the case of common law: date, time, country, entity, municipality and locality where the person was last seen, nationality, height, complexion, sex, age, description of particular signs, ethnicity, disability, and the agency that sent the information.

From 1968 to April 2018, the total number of registered missing persons was 37,435. However, it should be clear that, since then, with no public information available at the microdata level, to date there are 59,063 more missing persons! This contingent of more people could give a twist to the patterns shown below, but with the data at hand we have a scenario of the minimum possible. In the first instance, age and sex are considered as basic variables to describe the main characteristics of this missing population. The heatmaps suggest an increasing intensity of the phenomenon after the beginning of the so-called war against drug trafficking started in 2006. The graphs illustrate in qualitative terms that the higher the intensity of the color, the higher the number of disappearances (Fig. 1).

In absolute terms, the five states most affected by disappearances were: Tamaulipas (6131), State of Mexico (3918), Jalisco (3388), Sinaloa (3042), and Nuevo Leon (3042). It is worth noting that a particularly affected sector is the one made up of foreigners, mainly from the U.S. Among other findings, the main Mexican tourist destination of Quintana Roo was one of the most affected states in 2017. According to Table 2, for the male population, the state with the highest average age of missing persons is Tlaxcala (46.5) and for the female population it is Coahuila (28.2). The states with the lowest average age, excluding foreigners, are San Luis Potosí (28.8) and Tabasco (18.4), for men and women, respectively. The states with the highest standard deviation are Tlaxcala (22.8) and Mexico City (20.4). And the states with the lowest standard deviation are Baja California Sur (9.4) and Tabasco (6.4).

At the national level, of the total number of missing persons, 75% are men and 25% are women, with an average age of 31, 33, and 24 for the total, men, and women, respectively. On the other hand, the age range of this collective goes from 0 to 106 years old. A common aspect, which can be extracted from the source of information of common jurisdiction, is that the missing persons are slim and of

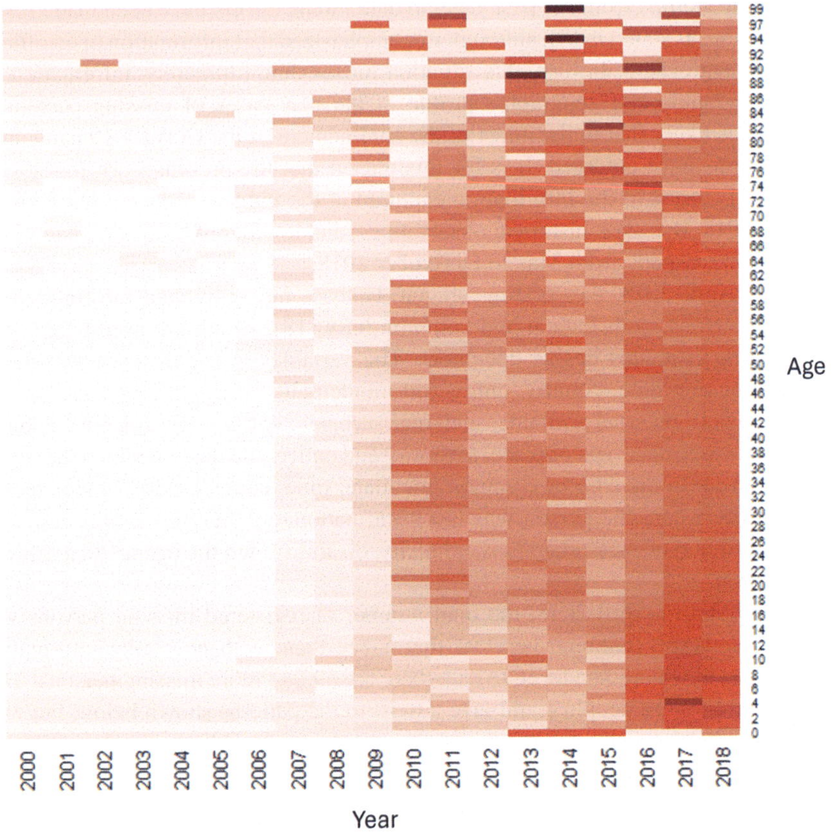

Fig. 1 Missing persons in Mexico, 2000–2018 (Source: Own elaboration with SEGOB data (2018))

average height. Likewise, 50% of the missing foreigners are under 27 years of age, 72% are male and come from the U.S. (63%). The demographic structure of missing persons is presented in contrast to those registered homicides [13]. The gender imbalance is notorious, as there is a high concentration of missing females between the ages of 10 and 15. In the male contingent, the modal age for both, missing and homicides, is around 30 years old. However, there is a large asymmetry for missing women. A possible explanation could be that they are victims of human trafficking (see [17]). It has been evidenced that one of the most significant consequences for the growth of homicides is a setback in life expectancy in Mexico [1] (Fig. 2).

3.2 Method

In Actuarial Sciences, it usually applied some smoothing method to appreciate the underlying trend in the mortality data, i.e., to achieve a mortality curve without

Table 2 Mean age (standard deviation) by state in Mexico, 2020

State	Men	Women	Total
Yucatán	44.3 (18.3)	23.3 (11.8)	37.6 (19.2)
Tlaxcala	46.5 (22.8)	20.5 (7.70)	35.9 (22.2)
Mexico City	42.1 (22.0)	27.8 (20.4)	35.9 (22.4)
Jalisco	34.3 (12.6)	27.9 (13.3)	33.6 (12.8)
Michoacán	34.1 (14.5)	26.4 (14.0)	32.6 (14.7)
Guanajuato	33.7 (13.1)	26.7 (15.6)	32.6 (13.8)
Coahuila	33.4 (13.2)	28.2 (13.9)	32.5 (13.5)
Chihuahua	33.4 (15.1)	22.4 (13.1)	32.2 (15.2)
Durango	33.5 (12.7)	25.0 (12.7)	32.2 (13.1)
Querétaro	35.0 (14.4)	25.7 (15.7)	32.1 (15.4)
Zacatecas	32.7 (13.9)	26.0 (11.9)	31.8 (13.8)
Baja California Sur	32.4 (9.40)	25.5 (14.8)	31.7 (10.1)
Nuevo León	33.5 (14.4)	26.3 (13.5)	31.6 (14.5)
Sinaloa	32.6 (12.3)	25.0 (12.9)	31.6 (12.6)
Tamaulipas	32.6 (12.8)	26.5 (12.6)	31.3 (13.0)
Nayarit	31.9 (10.4)	23.3 (7.60)	31.1 (10.5)
Campeche	37.3 (21.6)	18.8 (12.1)	30.2 (20.5)
Sonora	32.9 (14.1)	23.2 (12.9)	30.0 (14.5)
Guerrero	31.4 (13.0)	24.0 (12.9)	30.0 (13.3)
Veracruz	31.0 (12.5)	24.1 (12.7)	29.4 (12.9)
San Luis Potosí	28.8 (9.80)	25.7 (10.3)	28.4 (9.80)
Colima	31.2 (12.9)	22.3 (12.2)	28.4 (13.3)
State of Mexico	34 (19.10)	21.8 (14.1)	28.4 (18.0)
Oaxaca	34.0 (15.2)	20.4 (10.8)	27.8 (14.9)
Puebla	33.2 (19.3)	21.6 (14.2)	27.3 (17.9)
Morelos	30.5 (19.3)	21.5 (9.90)	27.0 (16.8)
Baja California	31.4 (16.2)	20.0 (12.0)	26.7 (15.6)
Quintana Roo	32.7 (17.0)	20.0 (12.0)	26.6 (16.0)
Hidalgo	30.0 (16.1)	21.4 (10.2)	26.2 (14.4)
Tabasco	31.9 (16.9)	18.4 (6.40)	25.7 (14.7)
Aguascalientes	30.4 (17.1)	20.4 (11.8)	25.5 (15.5)
Chiapas	28.8 (16.8)	18.6 (11.1)	24.1 (15.3)
Foreigner	25.2 (15.9)	14.8 (12.2)	21.0 (15.2)

Source: Own elaboration with SEGOB data (2018)

outliers useful for pricing. For this purpose, parametric models are an alternative, as is the controlled smoothing by segments of [12]. With that approach, it is desirable to maintain fidelity to the observed values when smoothing. When plotting the $lnqx's$, one can typically observe different patterns of variability in the log mortality data, which correspond to different life stages. Under these conditions, the proposed controlled smoothing is considered one of the best alternatives.

Some of the reasons for choosing the method of [12] are the following: it allows controlling the percentage of smoothness both in the whole curve and in

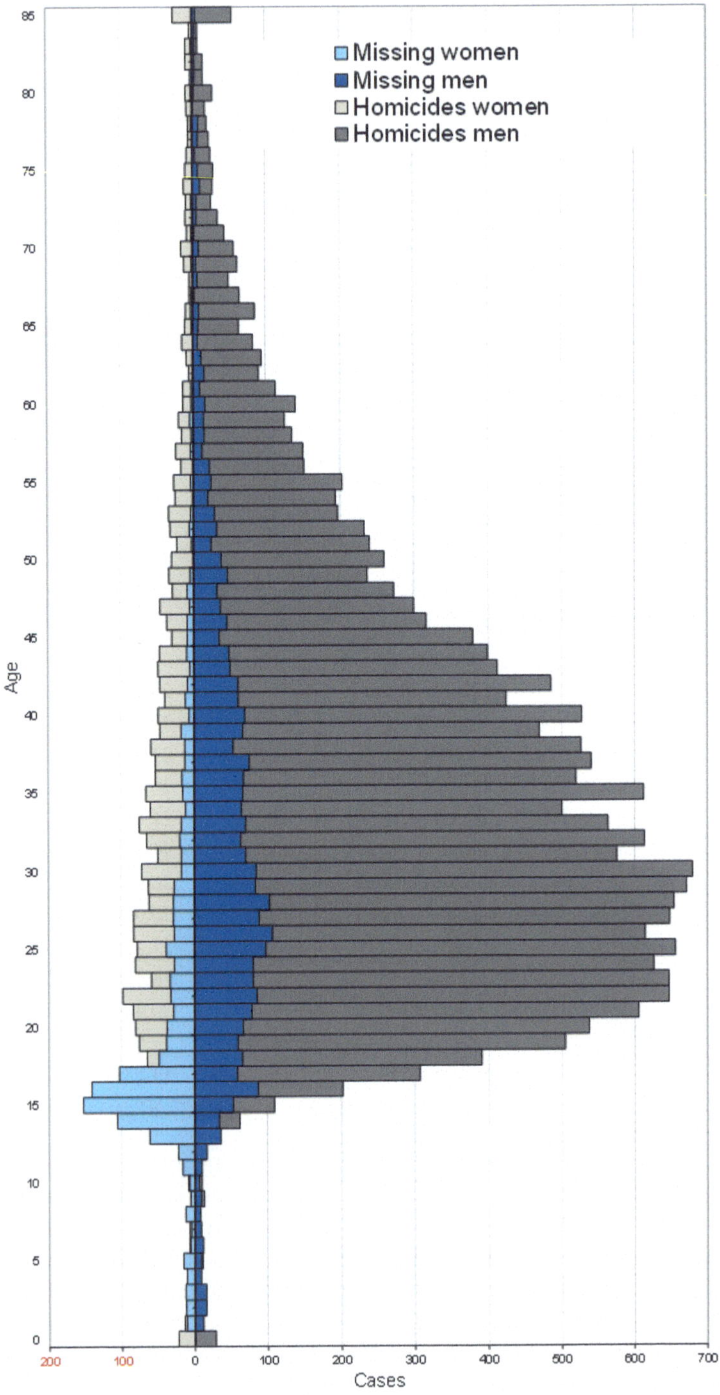

Fig. 2 Overmortality scenarios for women in collective insurance, 2018 (Source: Own elaboration)

the segments by means of smoothing indexes; the unions of the segments are continuous; intervals of the estimates can be generated; cutoff points of each segment can be set according to exogenous information; given two or more sets of mortality experiences, the analyst can set for all of them the same percentage of smoothness with the purpose of guaranteeing comparability in statistical terms. The reader interested in applying the method it is suggested to use the CSmoothing tool [19] freely available from https://anahuac.shinyapps.io/CSmoothing/. With this web tool, you can analyze your own data and access over 4,000 real mortality series from the Max Planck Institute for Demographic Research, the University of California, Berkeley, INED in Paris, and the United Nations.

The method is approached for two segments, without loss of generality, and the results are extended to the three-segment situations considered here. The function to be minimized (as with the Hodrick–Prescott filter) is the following:

$$Min_{\{\tau_i\}}\{\sum_{i=1}^{N_1}\frac{1}{\sigma_1^2}(y_i-\tau_i)^2+\sum_{i=N_1+1}^{N}\frac{1}{\sigma_2^2}(y_i-\tau_i)^2+\sum_{i=3}^{N}\frac{1}{\sigma_0^2}(\tau_i-2\tau_{i-1}+\tau_{i-2})^2\} \quad (1)$$

where σ_1^2 and σ_2^2 are the variances of the first and second data segments, with N_1 and $N_2 = N - N_1$ observations for each segment. Given the segmentation, it can be shown that the difference matrix, through which the smoothness is induced, can be written as follows:

$$K(i,j) = \frac{(-1)^{2+i-j}2!}{(j-i)!(2-j+i)!}$$

for i = 1, 2, . . ., N-2 and j = 1, 2,. . .,N, with $K(i,j) = 0$ if $j < i$ or $j > 2+i$. It is also shown in the method that the matrices that allow the continuity in the unions are

$$k_1 = \begin{pmatrix} 0_{2\times(N_1-2)} & \begin{matrix} 1 & -2 \\ 0 & 1 \end{matrix} \end{pmatrix} \text{ and } k_2 = \begin{pmatrix} \begin{matrix} 1 & 0 \\ -2 & 1 \end{matrix} & 0_{2\times(N_2-2)} \end{pmatrix}$$

with k_1 being of size $2 \times N_1$ and k_2 being of size $2 \times N_2$. Based on the aforementioned, in the segmentation we have the following matrix:

$$K = \begin{pmatrix} K_1 & 0 \\ k_1 & k_2 \\ 0 & K_2 \end{pmatrix} \quad (2)$$

where matrices K_1 and K_2 are of size $(N_1 - 2) \times N_1$ and $(N_2 - 2) \times N_2$ respectively. On the other hand, by means of Generalized Least Squares, it is obtained that the best unbiased estimate of the trend $\boldsymbol{\tau} = (\boldsymbol{\tau}_1, \boldsymbol{\tau}_2)'$, given by

$$\begin{pmatrix} \hat{\tau}_1 \\ \hat{\tau}_2 \end{pmatrix} = \left[\begin{pmatrix} I_{N_1} & 0 \\ 0 & I_{N_2} \end{pmatrix} + \begin{pmatrix} \lambda_1 I_{N_1} & 0 \\ 0 & \lambda_2 I_{N_2} \end{pmatrix} K^{*\prime} K^* \right]^{-1} \begin{pmatrix} y_1 \\ y_2 \end{pmatrix} \tag{3}$$

where I_{N_i} is the N_j-dimensional identity matrix and $\lambda_j = \sigma_j^2/\sigma_0^2$, for j=1, 2. Additionally, the variance–covariance matrix, used to estimate intervals, is given by

$$\Gamma = \text{Var}\begin{pmatrix} \hat{\tau}_1 \\ \hat{\tau}_2 \end{pmatrix} = \left[\begin{pmatrix} \sigma_1^{-2} I_{N_1} & 0 \\ 0 & \sigma_2^{-2} I_{N_2} \end{pmatrix} \right.$$

$$\left. + \sigma_0^{-2} \begin{pmatrix} K_1' K_1 + k_1' k_1 & k_1' k_2 \\ k_2' k_1 & K_2' K_2 + k_2' k_2 \end{pmatrix} \right]^{-1} \tag{4}$$

and unbiased estimators of the error variances are given by $\sigma_0^2 = \frac{RSS}{N-2}$ and $\sigma_j^2 = \lambda_j \sigma_0^2$ for $j = 1, 2$, with RSS the Residual Sum of Squares. The inverse of the variance–covariance matrix Γ can be understood as a precision matrix and can be written (for some weights $0 \leq c_i \leq 1, i = 1, \ldots, 4$) as a sum of precision matrices. It can be proven that the total precision accomplished by estimating the trend is the inverse of the variance–covariance matrix provided by Γ^{-1}. Therefore, the amount of smoothness to be attained globally by the trend for the unsegmented time series is $S = 1 - tr(I_N + \lambda K'K)^{-1}/N$. Guerrero and Silva [12] propose to measure the shared precision attributable to smoothness in each segment by means of the following statistics

$$S_j(\lambda_1, \lambda_2; N) = 1 - \frac{tr[B_{j,\lambda}(I_N + B_1 + B_2)^{-1}]}{N} \tag{5}$$

for j = 1, 2, that quantify the smoothness attained by smoothing segment j of the data, where

$$B_{1,\lambda} = B_1 B_0^{-1} = \frac{N_1}{N} \begin{pmatrix} \lambda_1 \left(\frac{N}{N_1} K_1' K_1 + k_1' k_1 \right) & \lambda_2 k_1' k_2 \\ \lambda_1 k_2' k_1 & \lambda_2 k_2' k_2 \end{pmatrix} \tag{6}$$

$$B_{2,\lambda} = B_2 B_0^{-1} = \frac{N_2}{N} \begin{pmatrix} \lambda_1 k_1' k_1 & \lambda_2 k_1' k_2 \\ \lambda_1 k_2' k_1 & \lambda_2 \left(\frac{N}{N_2} K_2' K_2 + k_2' k_2 \right) \end{pmatrix}. \tag{7}$$

They quantify the relative precision and have the following properties: (i) they take values between 0 and 1; (ii) they sum to unity; (iii) they are invariant under linear transformations; (iv) they have a linear behavior. A smoothness index per segment is deduced from each precision measure, which can be expressed as a percentage of smoothness attributable to each segment. The smoothness indices that the analyst controls by means of the smoothing parameter are bounded, which is why they can tend asymptotically to unity when the smoothing parameter tends to infinity. The same smoothness is required to make comparable estimates globally and by segments. As a basic principle, use a higher smoothing parameter when greater variability exists in a given data segment.

While this approach yields valuable insights, it does come with its set of limitations. Firstly, for a dependable implementation, the exclusion of missing values is paramount. Nevertheless [11] has illustrated that even when confronted with missing data, the Kalman filter can still be leveraged to derive estimates. Proficiency in advanced statistical techniques and programming is essential for effectively applying this filter. Secondly, a nuanced comprehension of the smoothness index and the appropriate cutoff points is crucial. To address this challenge, we suggest adopting criteria akin to those present in the works of Guerrero and Silva. Specifically, (a) higher variance in the data demands a correspondingly higher smoothness index, and (b) integrating exogenous information can assist in determining the cutoff points.

It is worth noting that proficient segmentation leads to a reduction in standard error within estimated trends, that is, in our case, mortality curves. An increase in error suggests that segmentation might not be necessary. In all applications of this chapter, segmentation consistently resulted in lower standard errors. To reestimate mortality, the data are taken from AMIS (2016–2018) and [18] for the years 2016–2018. Likewise, the remainder of 2018 was projected using exponential smoothing. The age and sex structure of missing persons was estimated according to [18] missing persons records. Then overmortality rates are calculated as follows

$$m_x = \frac{D_x + (\%K \times M_x) \times \%PMD}{P_x}$$

where D_x is the number of deaths at age x, $\%K$ is the percentage of missing persons with a life insurance policy, M_x is the missing population of age x assumed dead, $\%PMD$ is the percentage of such missing persons who were found lifeless, and P_x is the population of age x who are at risk of death, all in year t.

All estimations were performed using R statistical software version 4.3.1. Also, scenarios are provided for 2016, 2017, and 2018 by sex and for both individual and collective life insurance. The demographic structure of missing persons is assumed to be preserved by the end of 2018. For reference, the $lnqx's$ established by the Comisión Nacional de Seguros y Fianzas [7] are presented in each estimate. It is also assumed that death occurs in the same year in which the insured persons disappear, that homicides of missing persons are evenly distributed throughout the country, and that the probability of dying from homicide is the same for the entire population with or without life insurance.

Finally, to measure the impact of the excess mortality derived from the estimates by age segments, we use both life expectancy and temporary life expectancy proposed by Arriaga [4], between ages x and $x + i$, with and without the effect that a percentage of those who have died correspond to insured persons. This indicator is denoted by e(x,CNSF), referring to that coming from the CNSF, e(x) to the observed mortality, e(xM) to observed mortality plus the effect of missing persons under some assumptions, ie(x) to temporary life expectancy, and ie(xM) the same but adding the impact on missing people. The indicators are interpreted as the average number of years that a group of people alive at the exact age x will live from age x to age $x+i$ years, respectively.

4 Results

This section focuses on quantifying across all years, insurance types and sexes, what would happen to the respective life expectancies, at birth and over time, if two catastrophic scenarios are assumed: that $\%K = 100\%$, i.e. that the missing persons were insured, and that half of them and all of them were found alive, i. e. $\%PMD = 100\%$ or 50%. This can be considered an upper limit of what the Mexican insurance industry could expect in such circumstances in case of a massive claim by beneficiaries supported by a new regulation. While these scenarios may appear unrealistic, they serve to underscore the gravest potential challenges facing the Mexican insurance sector. By considering these possibilities, proactive measures can be taken to ensure the sector's economic resilience in the face of adversity. The respective life expectancy is added as a reference index based on the mortality rates approved by the CNSF.

Figures 3, 4, 5, and 6 first show the respective estimates for men by type of insurance and year. In the same order, the corresponding for women. By means of a shiny app developed for this work in the statistical software R version 4.3.1, the analyst can set the different smoothing parameters $(\lambda_1, \lambda_2, \lambda_3)$ and with them the smoothness, the cutoff points (N_1, N_2), the percentage of missing persons with a life insurance policy $(\%K)$, and the percentage of such missing persons who were found lifeless $(\%PMD)$. Mortality experiences are taken from age 12 to 85 years and older, for both sexes. For all estimates, the following parameters are taken: overall smoothness S% = 75% (as suggested in [12]), for the first segment $S_1\% = 70\%$, and for the following $S_2\% = 75\%$ and $S_3\% = 77\%$ respectively, which are obtained by setting the smoothing parameters $\lambda_1 = 3.9$, $\lambda_2 = 5.0$ and $\lambda_3 = 9.2$, with cutoff points $N_1 = 3$ (i.e., age 18), and $N_2 = 18$ (i.e., age 30).

In Table 3, in general, it can be seen how life expectancy undergoes changes through the implementation of such assumptions. It is important to note that the mortality rates m_x, with which the indicator is estimated, are calculated in a similar way for both the male and female populations, as well as the fact that the source of information is the AMIS for deaths (claims) and exposed by type of insurance, respectively. For the male case and individual insurance, among other things, life expectancies at birth for the 3 years, in all scenarios, exceed the life expectancy expected with the experience of the CNSF. In other words, in this case the Mexican insurance industry would have considered a higher mortality than was actually observed, even including the effect of adding the missing persons. For the same case, but with collective insurance, lower life expectancies are observed for the years 2016 and 2018, i.e., the mortality curve expected by the CNSF is systematically

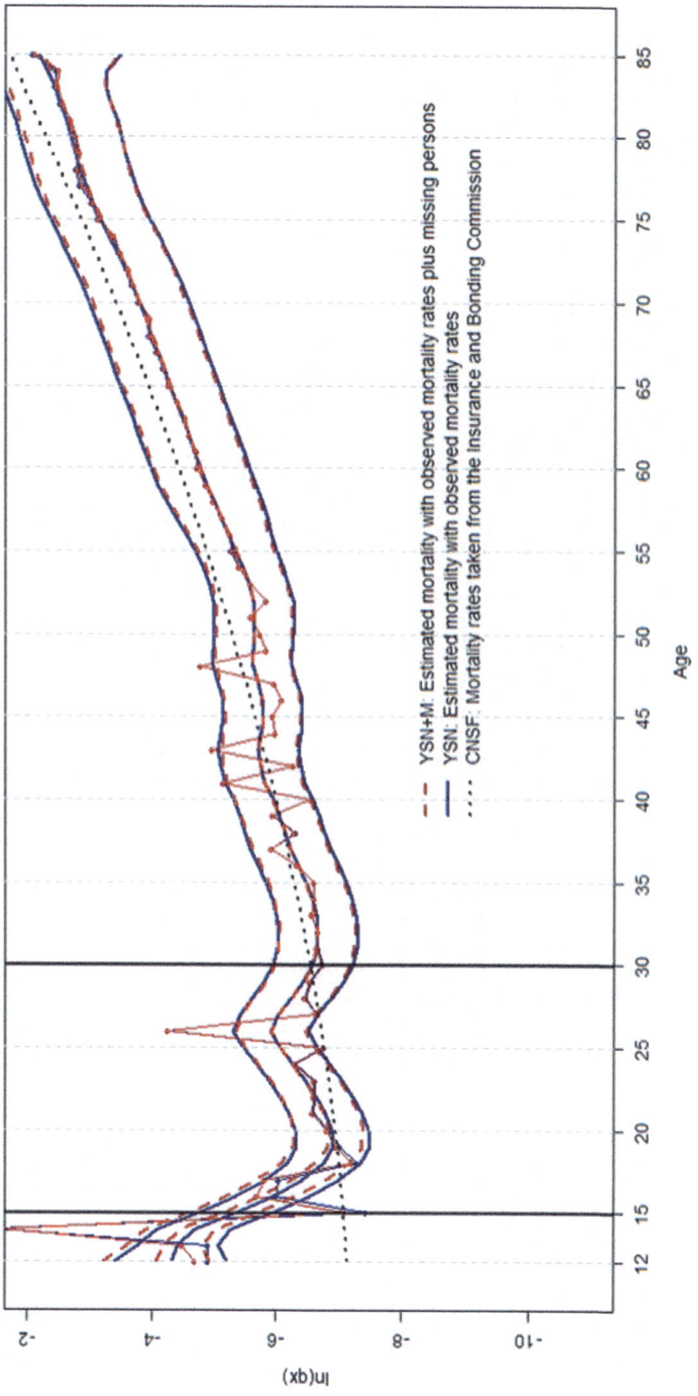

Fig. 3 Overmortality scenario for men with $\%PMD = 50\%$ in collective insurance, 2018 (Source: Own elaboration)

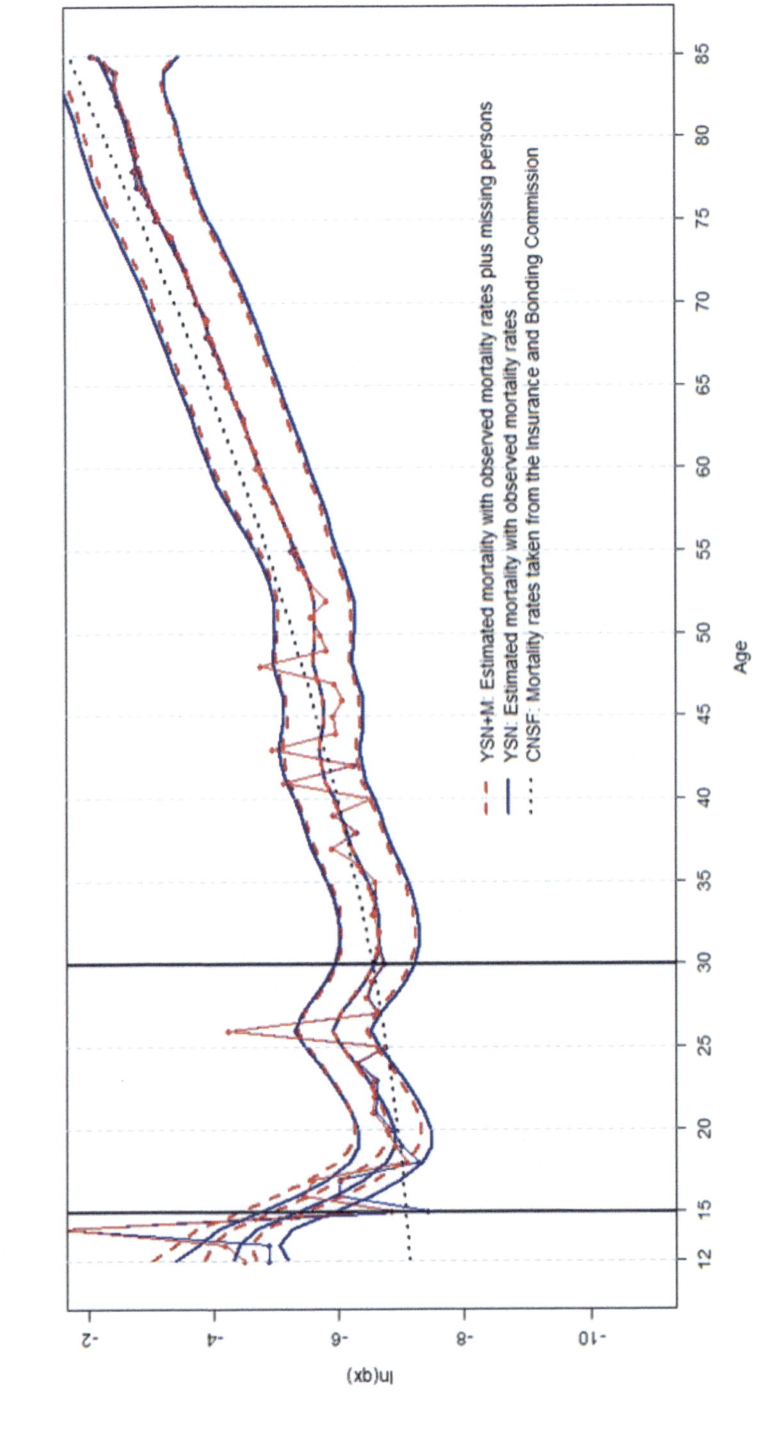

Fig. 4 Overmortality scenario for men with $\%PMD = 100\%$ in collective insurance, 2018 (Source: Own elaboration)

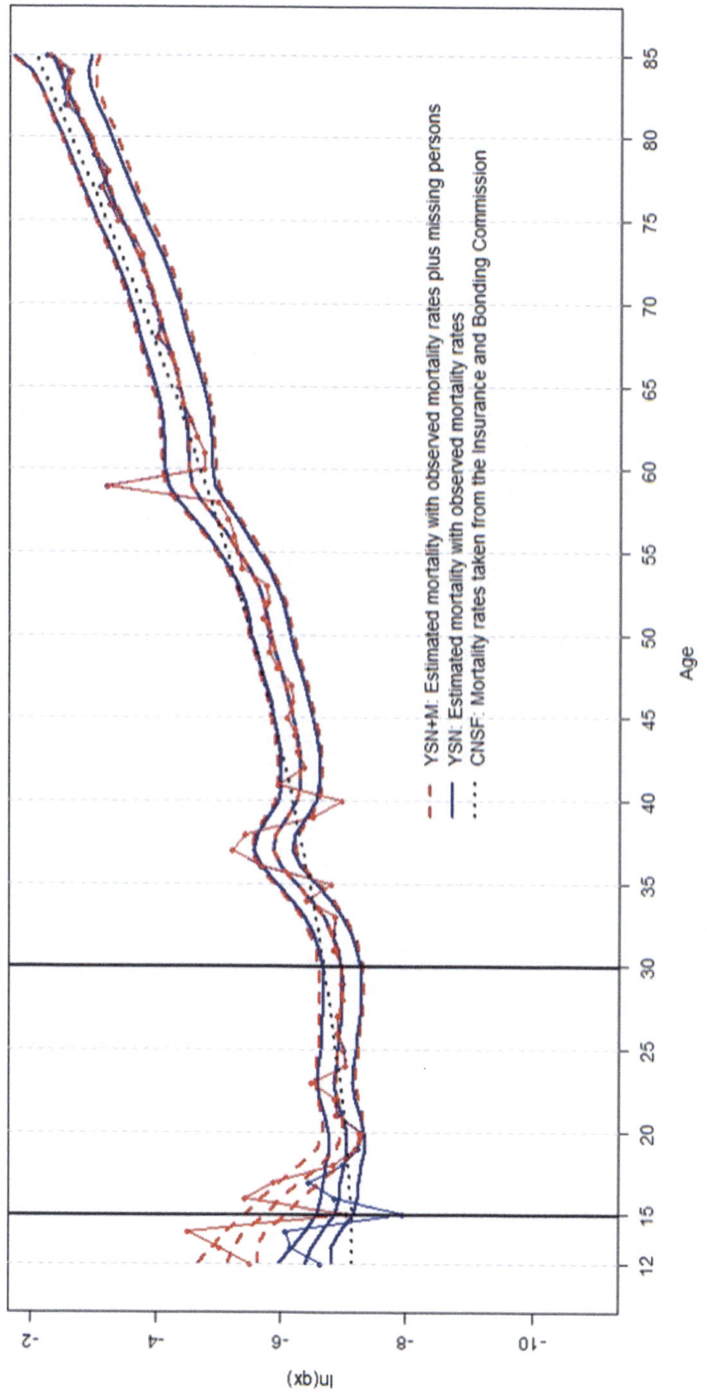

Fig. 5 Overmortality scenario for women with $\%PMD = 50\%$ in collective insurance, 2018 (Source: Own elaboration)

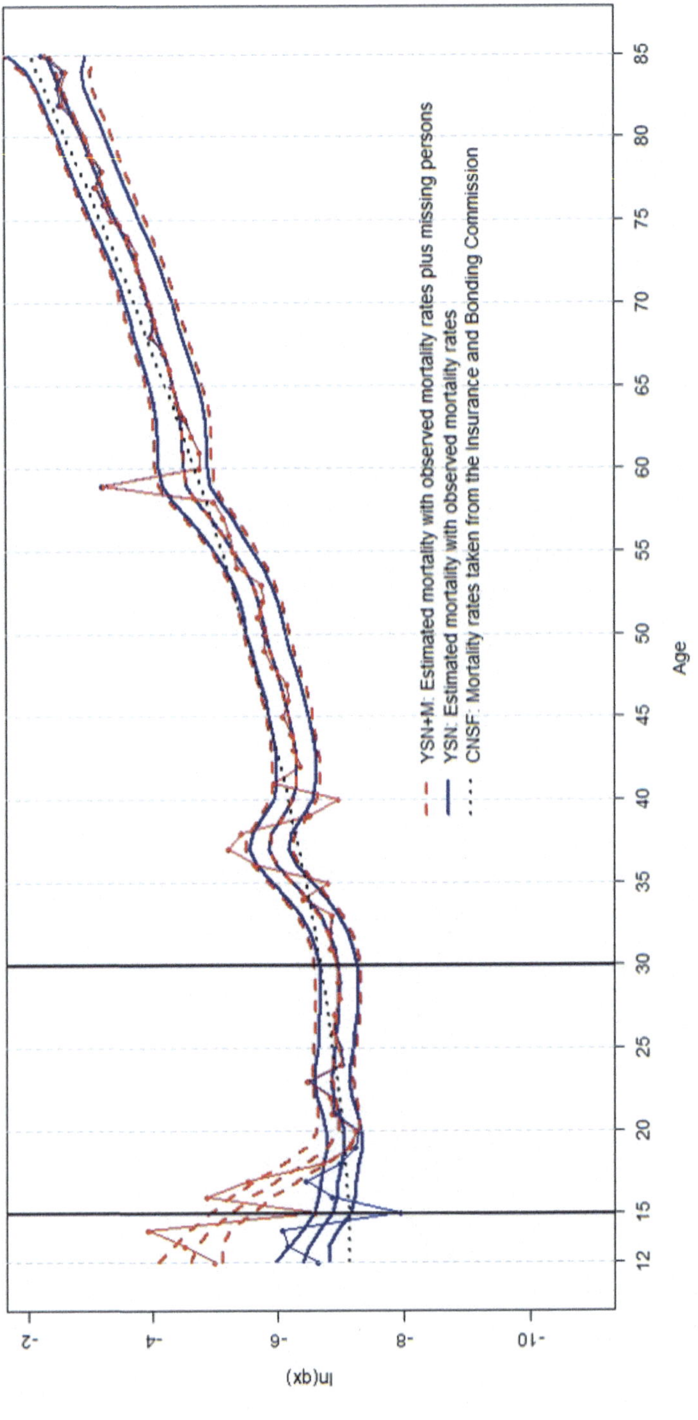

Fig. 6 Overmortality scenario for women with $\%PMD = 100\%$ in collective insurance, 2018 (Source: Own elaboration)

Table 3 Catastrophic scenarios and life expectancies by insurance type, sex, and year, 2016–2018

K%	PMD%	Insurance-year	e(x,CNSF)	e(x)	e(xM)	ie(x)	ie(xM)
Men							
100	50	Individual-2016	56.63	67.42	66.99	14.88	14.88
100	100	.	.	.	66.59	.	14.87
100	50	2017	56.63	67.51	66.97	14.89	14.88
100	100	.	.	.	66.46	.	14.87
100	50	2018	56.63	67.59	67.06	14.88	14.87
100	100	.	.	.	66.54	.	14.86
100	50	Collective-2016	61.15	61.51	61.23	14.86	14.86
100	100	.	.	.	60.91	.	14.86
100	50	2017	61.15	63.49	63.34	14.85	14.85
100	100	.	.	.	63.15	.	14.85
100	50	2018	61.15	60.7	60.05	14.81	14.81
100	100	.	.	.	59.33	.	14.81
Women							
100	50	Individual-2016	62.58	69.33	68.74	14.95	14.94
100	100	.	.	.	68.16	.	14.94
100	50	2017	62.58	69.43	68.65	14.95	14.95
100	100	.	.	.	67.93	.	14.95
100	50	2018	62.58	69.4	68.69	14.94	14.94
100	100	.	.	.	68.04	.	14.94
100	50	Collective-2016	63.25	64.01	63.60	14.87	14.87
100	100	.	.	.	63.17	.	14.87
100	50	2017	63.25	66.65	66.36	14.91	14.91
100	100	.	.	.	66.06	.	14.91
100	50	2018	63.25	64.04	63.11	14.85	14.85
100	100	.	.	.	62.16	.	14.85

Source: Own elaboration

exceeded, even without adding the missing persons. In fact, the most alarming case would be for the last year as there would be a difference of almost 2 years. The observed temporary life expectancy compared to that including the missing persons is practically the same in each case.

For the female case with individual insurance, similar to the male case, it can be observed that, in the extreme scenario, considering missing women, life expectancies are higher than those regulated by the CNSF. In other words, the Mexican insurance sector is considering a higher mortality for all years. For the same group in collective insurance, the lowest life expectancy is observed in 2016 in the extreme case, and in 2018 in both scenarios. A maximum discrepancy of just over 1 year can be seen in 2018. In this insurance, the temporary life expectancies show no changes. The last figure shows the worst-case scenarios by sex and insurance type, which are the same for 2018.

5 Conclusions

The assumption that the missing policyholders have a probability that half or all of them will turn up deceased, i.e., a couple of not implausible assumptions, is considered useful to be able to estimate scenarios with respect to the real mortality experienced in the Mexican context. This is relevant in view of an eventual obligation on the part of the insurance sector to cover insured sums through the so-called death certificates. At the micro-data level, it can be stated that the main effects, measured as lower life expectancy than expected and regulated by the CNSF, occur in collective insurance. It is therefore urgent to estimate this risk, given that, in some cases, the pure measurement of the risk without even considering the missing persons is greater than expected.

On the other hand, the controlled smoothing method used is thought to be a useful technique for segmenting mortality and estimating its trends for comparisons among years. Thus, it is believed that the approach presented, even though it is an exploratory statistical tool, is sufficient to demonstrate the need to rethink the estimation of new mortality experiences, mainly for collective insurance for both sexes. Controlled smoothing by segments also provides versatility to estimate trends with different degrees of variability for data segments defined by the analyst. An added value of the method is the continuity achieved in the mortality curves, despite not having the same smoothing levels and the fact that there are no distributional assumptions to be verified.

References

1. Aburto, J., Beltrán, H., García, V. y Canudas, V. (2016). Homicides in Mexico reversed life expectancy gains for men and slowed them for women, 2000–10. Health Affairs, 35: 88–95.
2. Arista, L. (2021). Los desaparecidos, la otra pandemia que azota a México. Expansión/polìtica. Available at https://politica.expansion.mx/mexico/2021/08/06/voces-los-desaparecidos-es-la-otra-pandemia-que-azota-a-mexico
3. Arteta, I. (2021) Base de datos de desaparecidos 2011–2020, en formato de datos abiertos. Animal político. Available at https://www.animalpolitico.com/2021/07/base-de-datos-desaparecidos-2011-2020-datos-abiertos/
4. Arriaga, E. (1984) Measuring and Explaining the Change in Life Expectancies. Demography, 21(1): 83–96. doi: 10.2307/2061029
5. AMIS (2013-2020) Reportes de Sistema Estadístico del Sector Asegurador (SESA) de Vida y Pensiones. Available at https://centroestadisticoamis.mx/vida-reportes-sesa/.
6. CNB (2022) Contexto General de personas desaparecidas. Available at here.
7. CNSF (2018) Anuario Estadístico de Seguros y Fianzas 2018. Available at here.
8. Cornelius, N. (2019) Life Insurance Claims for a Missing Person. Quotacy. Available at https://www.quotacy.com/life-insurance-claims-for-a-missing-person/
9. Dubey, N. (2021) How to get a claim when a policy buyer goes missing. Mint.here.
10. Fuentes, M. L., y Arellano, S. (2021) El desarrollo frente a la pandemia: nuevos escenarios, nuevos retos. En Informe del Desarrollo en México, Coordenadas para el debate del desarrollo, p. 89. UNAM.
11. Guerrero, V. M. (2008). Estimating trends with percentage of smoothness chosen by the user. International Statistical Review, 76(2), 187–202. Available at here.

12. Guerrero, V. M. y Silva, E. (2015). Smoothing a Time Series by Segments of the Data Range. Communications in Statistics - Theory and Methods, 44:21, 4568–4585. Available at here.
13. INEGI (2013–2020) Estadísticas Vitales de mortalidad: homicidios. Dirección General de Estadística. Available at https://www.inegi.org.mx/programas/mortalidad/
14. INEGI (2020) Censo de Población y Vivienda 2020. Dirección General de Estadística. Available at https://www.inegi.org.mx/programas/ccpv/2020/
15. Kang, Y., Gao, S., Liang, Y., Li, M., Rao, J., and Kruse, J. (2020). Multiscale dynamic human mobility flow dataset in the US during the COVID-19 epidemic. Scientific data, 7(1), 1–13.
16. Lambertucci, C. (26 agosto 2021). Aumento de la violencia y falta de peritos: la crisis forense en México deja más de 52.000 cuerpos sin identificar. El país. Available here.
17. López Marroquín, S. (2019). Desaparición, esclavitud y trata de personas: situación de las mujeres en México. Cuicuilco. Revista de ciencias antropológicas, 26(74), 163-181.
18. SEGOB (2018) Registro Nacional de Datos de Personas Extraviadas o Desaparecidas (RNPED). Secretariado Ejecutivo del Sistema Nacional de Seguridad Pública.gob.mx. Available at here.
19. Silva, E., Peralta A. y Peralta, E. (2021) Exceso de mortalidad en México 2020: una estimación preliminar a nivel nacional y estatal, Población y Salud en Mesoamérica.
20. Torres, Y. (2016) Cómo cobrar el seguro de una persona desaparecida?. El Financiero. Available at here.
21. Torres, Y. (2014) Difícil, el cobro de seguros de desaparecidos. El Financiero. Available at here.
22. U.S. Census Bureau (2020) Population and Housing State Data, Population Division. Available at here.

Development of Cyclical Indicators Based on Multivariate Spectral Analysis

Alfredo Bustos and Noemi López-Delgado

Abstract In this chapter, we review a method proposed by Bustos (Statistical Methodology for leading Indicator, Bulletin of the International Statistical Institute. Contributed papers, Book I. 49th session Florence, Italy, 1993, pp 189–190) to develop sets of optimal weights for constructing cyclical indicators. To achieve this, we employ the canonical analysis of multivariate time series in the frequency domain, as described by Brillinger (Time Series, Data Analysis and Theory, Holden Day, Inc., San Fran-cisco; 1981). For comparative purposes, we use the sets of coincident and leading indicators utilized by the Cyclical Indicators System of the National Institute of Statistics and Geography (SIC-INEGI) from January 2004 to March 2020. The initial application of the proposed procedure to this dataset does not yield optimal results. However, we demonstrate how our approach allows for evaluating candidate indicators for inclusion in the analysis, resulting in refined sets. Following this selection process, a new application of the methodology is conducted. Considering criteria such as cross-correlation at different lags and the ability to forecast the coincident indicator from the leading one, we assess the improvements offered by our proposal. The results obtained through this approach surpass the performance of indicators developed using traditional methods in several aspects.

Keywords Composite indicators · Cyclical · Coincident · Leading · Cross correlation function · Multivariate time series · Spectral analysis · Frequency domain · Canonical correlation · Forecast

---

A. Bustos · N. López-Delgado (✉)
Instituto Nacional de Estadística y Geografía, Aguascalientes, Mexico
e-mail: alfredo.bustos@inegi.org.mx; nohemi.delgado@inegi.org.mx

J. A. Christen et al. (eds.), *Statistics, Society and Environment*, Springer Proceedings in Mathematics & Statistics 479, https://doi.org/10.1007/978-3-031-78401-9_9

1 Introduction

Human beings have developed ingenious methods to anticipate the future. Among these are systems of coincident composite indicators (CCI) and leading indicators (ICA), designed to timely identify variations inherent in the cyclical behavior of an economy. Despite being developed almost a century ago, the basic methodology of these systems has undergone only minor changes. Although this might not have been explicitly stated, we believe the primary purpose of cyclical indicator systems (SIC) is to obtain accurate forecasts of future values of the coincident composite indicator based on the current information used for calculating the ICA.

The methodologies currently used for cyclical indicator systems (SIC) originated from the pioneering work of Mitchell and Burns at the National Bureau of Economic Research (NBER) in the United States during the early twentieth century. Following in their footsteps, their disciple Geoffrey Moore has contributed to improving these methods, along with various national and international organizations. However, the fundamental aspects of the original proposal remain intact across all implementations.

Given that no single indicator can adequately summarize the recent cyclical behavior of a country's economy, efforts are made to identify a set of indicators, termed coincident indicators, whose aggregation is expected to approximate this behavior better. Typically, this identification involves dating changes in each indicator's evolution direction and comparing these dates with those of a reference indicator that reflects the aggregate production behavior in the productive sectors. When directional changes in peaks or troughs frequently coincide, the indicator is considered coincident. Similarly, we identified leading indicators, whose peaks and valleys generally precede those of the reference indicator.

The absence of an optimality criterion to justify the methodology led to its abandonment due to the difficulty in comparing proposals and results. Nowadays, even without such a criterion, there seems to be a consensus that weightings should be based on the variability of each indicator. This approach aims to avoid giving greater weight to those indicators that exhibit greater instability during the observation period.

A recent addition to the methodology involves using versions of indicators with reduced or eliminated secular trends. This includes methods such as the Hodrick–Prescott filter [3] or as employed by the Organization for Economic Cooperation and Development [5], spectral analysis of time series. In the latter case, the approach involves reducing the power of one or more lower frequencies using a filter.

The experience with methodologies like those summarized in the previous paragraphs is rather mixed. Since a key objective is to anticipate undesirable declines in economic performance to mitigate negative effects on people and businesses, it seems sensible to choose the option that results in the lowest forecast error. In Mexico's case, INEGI decided to change its methodology, which it had followed until 2009, based on this experience (see [4, 9]).

Throughout the remainder of this section, we will address what we consider to be the most important limitations of the methodologies currently in use.

Incorporating the theoretical and methodological advancements of multivariate time series statistical analysis does not seem customary, nor does the use of current information technologies. Additionally, there is a notable absence of an optimality criterion to establish more accurate values for the coefficients of the indicators.

The proposal summarized by Bustos [2] addresses several of the aforementioned and limitations concerns. His main source of inspiration is Brillinger ([1], Ch. 10), specifically his treatment of canonical time series analysis, leveraging the frequency domain perspective. The basic reasoning behind this proposal is that, when studying the business or economic cycle, it is natural to approximate the time series representing it through weighted sums of periodic functions with stochastic coefficients over a range of frequencies, that is, their Fourier transforms. Recent applications of this approach to econometric analysis can be found in Pollock ([6, 7], and [8]).

This approach offers several key advantages. Firstly, this approach avoids the loss of information that can occur with other methods. Instead, it provides a valuable additional perspective through a complementary analysis. Also, in the stationary case, the second moments are simplified, allowing for the transfer of some multivariate techniques from the statistical analysis of random vectors and their application frequency by frequency. Similarly, the common secular trend across all series can be removed by reducing the power in the low frequencies of the spectrum of the first principal component, obtained from all the series under consideration. Stationary versions of the original indicators are derived from the modified set of principal components.

Another consequence of this approach is the introduction of explicit criteria to evaluate the quality of the results, which differs from customary practices. For instance, the canonical analysis aims to identify vectors of coefficients that maximize the square of the norm of the coherences at each frequency. It is reasonable to assume that, returning to the time domain, the results will also exhibit optimal cross-correlations for one or more lags. Consequently, the forecasts provided by the ICA for the CCI are expected to be optimal in some sense.

Another limitation that this proposal sought to address is the unverified assumption that a single composite indicator for each set summarizes all the cycle-related information contained in the original data. By allowing for more than one set of pairs, the suggested approach enables us to identify the optimal reduction of dimensions, ensuring a more comprehensive and accurate representation of the data.

The remainder of this chapter will proceed as follows. The following section presents an overview of the proposed methodology and some relevant results. We then apply our proposal to the information used in constructing the SIC-INEGI, allowing for a comparison between both sets of results and introducing the criteria we consider appropriate. Subsequently, we propose a strategy to eliminate undesirable indicators from the set of coincident ones, and a similar approach will be taken with the leading indicators. Finally, using the results of the previous selection process, we apply our methodology anew and compare its outcomes with those of the current INEGI procedure. We conclude with a summary of our results and discuss some future research approaches to leverage the available information fully.

2 Proposed Methodology

Please refer to Fig. 1 for an overview of the complete proposal for calculating a system of cyclical indicators. The analysis is divided into two parts. First, we perform a principal component analysis in the frequency domain on the original information set of the n series $\{\underline{X}_t\}_{t=1}^{T}$ and their second-moment estimators $\{\Gamma_X(k)\}_{k=0}^{T-1}$, if circumstances require it.

In all cases, their finite and discrete Fourier transforms, specifically the vector $\underline{v}_X(\lambda)$ and the matrix $\Phi_X(\lambda)$, are located in the frequency domain. For each frequency value, $\lambda_j = 2\pi j/T$, $j = 0, \ldots, T-1$. The principal component analysis is then carried out, starting from the diagonalization of the Hermitian spectral matrix, $\Phi_X(\lambda_j)$. Let matrix B whose columns are the characteristic vectors of the spectral matrix provide a new set of Fourier coefficients. This results in a diagonal spectral matrix. The inverse Fourier transform generates a new set of series, $\{\underline{Y}_t\}_{t=1}^{T}$, and its corresponding covariance structure, $\{\Gamma_Y(k)\}_{k=0}^{T-1}$, which is approximately formed by diagonal matrices. This approach ensures that the new set of series exhibits zero cross-correlations, aligning with the analysis of principal components of independent random vectors.

The principal component analysis enables us to minimize the impact of specific frequencies in new iterations of the original series. When these frequencies align with the lowest ones, secular trends become less pronounced. This approach aligns with the methodology set forth by the OECD, with the key distinction being that their method involves the elimination of low frequencies on a series-by-series basis.

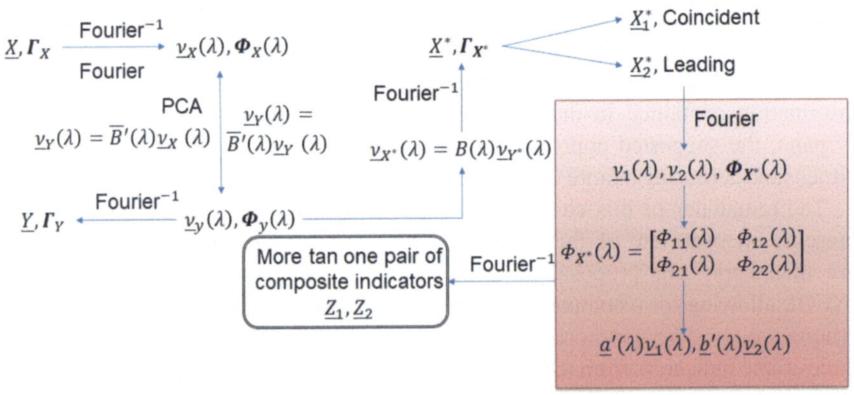

Fig. 1 Calculation process for pairs of coincident and leading indicators

The canonical analysis proceeds from the original set of series or from a set where trends have been attenuated, as aforementioned. In this case, to make efficient use of all available information, the definition of the sets of coincident or leading series considers the behavior of the cross-correlation structures of each candidate series with a reference indicator. In short, this approach excludes the consideration of turning points at peaks and troughs and their dating.

Once the sizes of each set have been determined (denoted by p_1 and p_2), the spectral matrix of the vector whose first p_1 components are formed by the coincident series is estimated, and the series advanced are occupied by the remaining (p_2). The estimation is carried out as indicated in (1):

$$\Phi_X(\lambda) = \sum_{j=-P}^{P} \frac{P-|j|}{P} \hat{\Gamma}_j e^{-i\lambda j} \tag{1}$$

$$\Phi_X(\lambda) = \begin{pmatrix} \Phi_{11} & \Phi_{12} \\ \Phi_{21} & \Phi_{22} \end{pmatrix} \tag{2}$$

Based on this information, we can pose the optimization problems indicated in (3) for each frequency. The solutions to these problems will yield the coefficient vectors $\underline{a}(\lambda_j)$, $\underline{b}(\lambda_j)$ that lead to the canonical spectral pairs (see [1], Theorem 10.3.2).

$$\begin{aligned} \max_{\underline{a}(\lambda_j),\underline{b}(\lambda_j)} &= |Corr\{\underline{a}'(\lambda_j)\underline{v}_1(\lambda_j)\underline{b}'(\lambda_j)\underline{v}_2(\lambda_j)\}| \\ &= \left| \frac{\underline{a}(\lambda)\Phi_{12}\underline{b}(\lambda)}{\sqrt{\left(\underline{a}'(\lambda)\Phi_{12}\underline{a}(\lambda)\right)\left(\underline{b}'(\lambda)\Phi_{12}\underline{b}(\lambda)\right)}} \right|^2 \end{aligned} \tag{3}$$

We have that $\underline{a}(\lambda_j)$ and $\underline{b}(\lambda_j)$ maximize the squared norm of the coherence within the canonical pair. It is possible to obtain new pairs of vectors by imposing null correlation conditions between canonical pairs to solve optimization problems similar to (3). This process can be carried out so often as the smaller value between p_1 and p_2. Furthermore, under the identifiability condition that the coefficient vectors obtained in this way have a norm equal to 1, these vectors are nonzero solutions to the systems of equations presented in (4). In other words, the vectors $\underline{a}(\lambda_j)$ and $\underline{b}(\lambda_j)$ are eigenvectors of the matrices that appear in expression (4). It can be proven that the corresponding eigenvalues $\rho(\lambda)$ coincide with the optimal value of the maximization criterion presented in (3).

$$\left.\begin{aligned} \Phi_{11}^{-1}(\lambda)\Phi_{12}(\lambda)\Phi_{22}^{-1}(\lambda)\Phi_{21}(\lambda) &= \rho_j(\lambda)\underline{a}_j(\lambda) \\ \Phi_{22}^{-1}(\lambda)\Phi_{21}(\lambda)\Phi_{11}^{-1}(\lambda)\Phi_{12}(\lambda) &= \rho_j(\lambda)\underline{b}_j(\lambda) \end{aligned}\right\} j = 1, \ldots, \min\{p_1, p_2\} \tag{4}$$

From each of these solutions, new sets of Fourier coefficients are defined. Once more, the inverse Fourier transform of these results will generate a new series of indicators, with the highest cross-correlation occurring at a specific lag. This is true for any other series obtained through a similar process. A new series vector, $\{\underline{Z}_t\}_{t=1}^{T}$, is formed with the coincident indicators concentrated in the first p_1 components and the leading indicators in the subsequent p_2 components. By construction, the resulting indicators can be calculated using contemporary and lagged values of the original series. In this way, the implicit restrictions in the calculation of indices, which assume that the values of the coefficients for these lags are equal to zero, are no longer valid. Therefore, it is reasonable to think that the optimality criterion will achieve better levels.

3 Numerical Example

To illustrate the concepts presented in the previous chapter, we will analyze the series published in the Economic Information Bank (BIE15), comprising 195 observations from January 2004 to March 2020, accessible on the INEGI website.

To ensure a fair comparison, we will analyze first the results obtained through INEGI's methodology. Figure 2 illustrates both the coincident and leading indicators calculated using this methodology.

The blue line represents the coincident composite index, and the red line represents the leading composite index. Near the peaks and troughs, the behavior is satisfactory before 2013. However, after 2013, the relationship becomes somewhat unclear. The leading indicator rises while the coincident indicator appears to decline, and shortly before reaching 2020, the coincident indicator falls before the leading

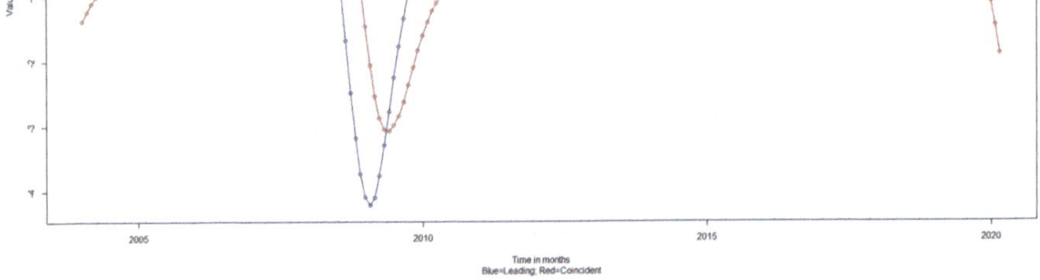

Fig. 2 Leading and coincident indicators, SIC-INEGI

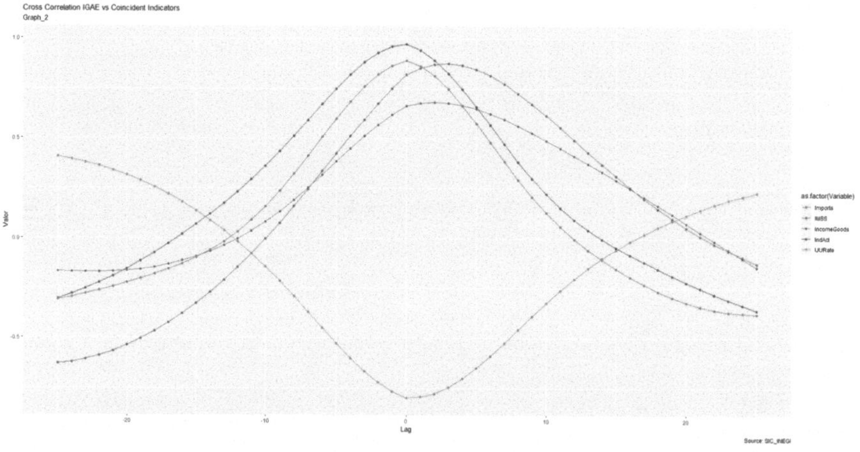

Fig. 3 Cross correlation IGAE vs each coincident indicator

one. The cross-correlation structure between the two indices peaks at 0.7724 with a lag of 5 months. From our perspective, identifying an indicator as coincident or leading based solely on the dating of turning points and comparing these dates with a reference series ignores information on the relationship between the two series throughout the entire observation interval. Using the cross-correlation patterns between both series makes more efficient use of all available information.

Figure 3 indicates that the series of Industrial Activity, Imports, and the Urban Unemployment Rate align with the IGAE, as the maximum correlation with the IGAE occurs at lag zero. In contrast, the registration of insured people in the IMSS and income from the supply of goods and services appear to lag. Industrial activity has the highest correlation (0.9736), while income from the supply of goods and services has the lowest correlation (approximately 0.7) with an apparent lag of 1, 2, and 3 months.

A comparison of the same IGAE against the leading indicators (Fig. 4) reveals that the Stock Market Price and Quotations Index, the Standard and Poor's 500 Index, the Business Confidence Indicator, and the Employment Trend in Manufacturing lead the IGAE, with the maximum values of the correlations occurring at negative lags. The Stock Market Price and Quotations Index has the highest correlation (0.8719). Conversely, the Equilibrium Interbank Interest Rate (TIIE) appears to be a lagging indicator though with a weak correlation (approximately 0.4), and the Mexico–US Bilateral Real Exchange Rate shows a behavior more aligned with coincident indicators, with a correlation close to -0.3, the lowest in this group of indicators.

Figure 5 illustrates the squared norm of the coherence for three out of six canonical pairs across all harmonic frequencies $0 \leq \lambda < 2\pi$, and for the pair comprising the coincident and leading indicators of the SIC. Indeed, the first canonical pair is optimal, but some of the remaining canonical pairs could also provide useful

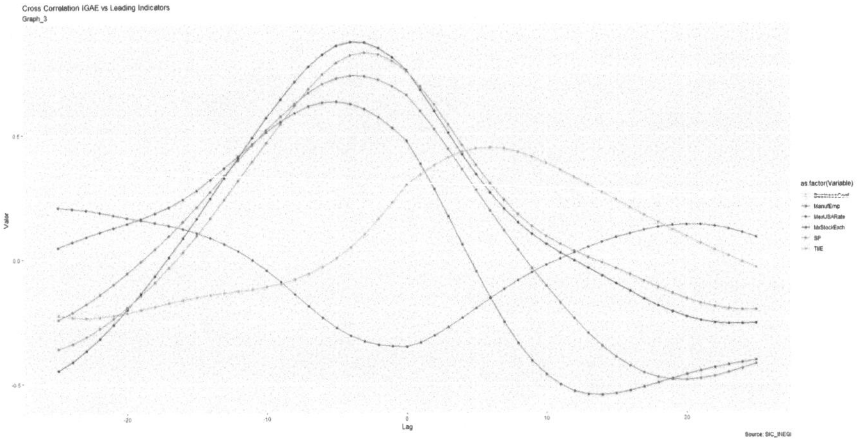

Fig. 4 Cross correlation IGAE vs. each leading indicator

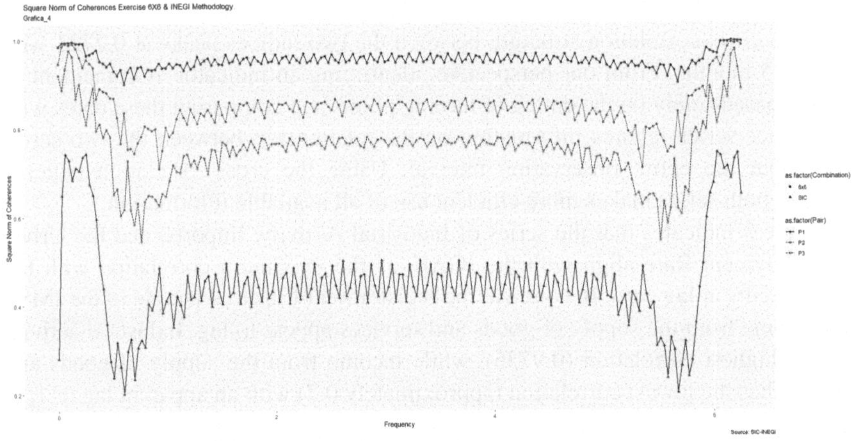

Fig. 5 Eigenvalues for the first three canonical pairs and SIC

information. By construction, these pairs are independent of the first pair (almost orthogonal), which means it is unnecessary to restrict oneself to a single pair of indicators. Figures 6 and 7 illustrate respectively the first and second canonical pairs resulting from the proposed approach. While the relationship between coincident and leading indicators in the first canonical pair is not immediately evident, the second pair demonstrates a more pronounced correlation between the two indices.

Figure 8 indicates that the second canonical pair exhibits slightly superior performance than the first in terms of maximum cross-correlation values. The first canonical pair leads approximately 5 months with a maximum correlation of 0.69. However, the second canonical pair has a maximum correlation of 0.93, but it does not appear to lead. About the SIC, there is a maximum cross-correlation between

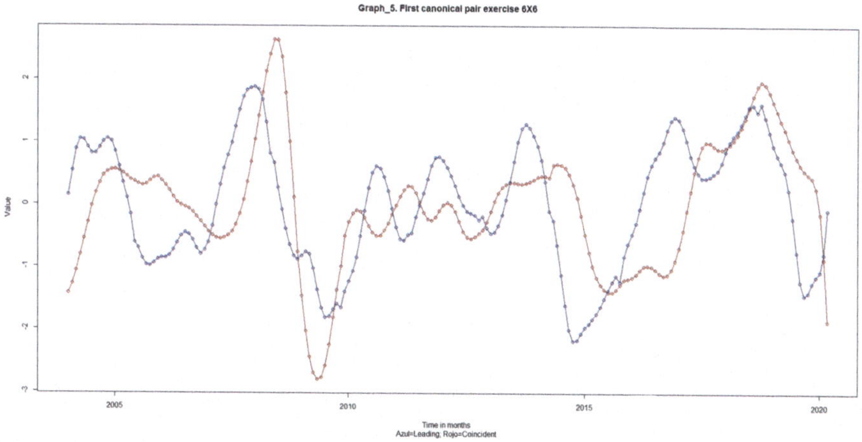

Fig. 6 First canonical pair, 6 coincident and 6 leading indicators (6 × 6)

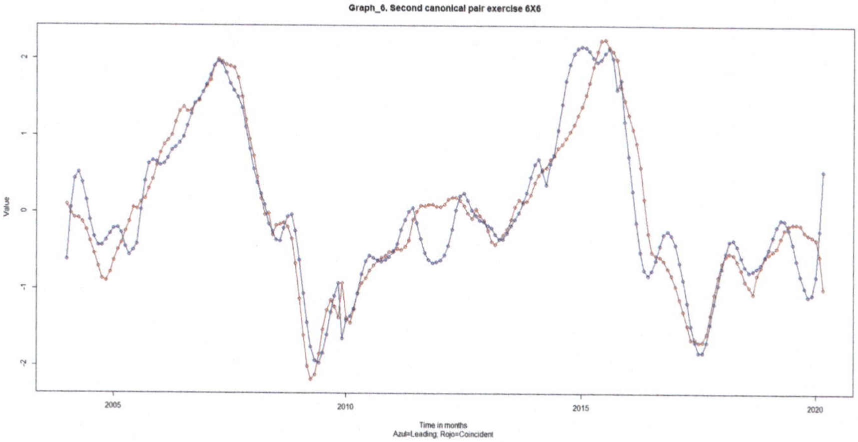

Fig. 7 Second canonical pair, 6 × 6

coincident and leading of 0.79, for a 5-month lead. This suggests that optimality in the frequency domain may not necessarily be reflected in the time domain.

3.1 *Results of the Comparison Proposal*

In light of these unsatisfactory outcomes, we forward a few hypotheses to address the situation. First, we recommended that our input, namely the series, be subjected to a more thorough examination. Indeed, no previous statistical analysis was conducted to evaluate the quality of the information. Our initial analysis, based

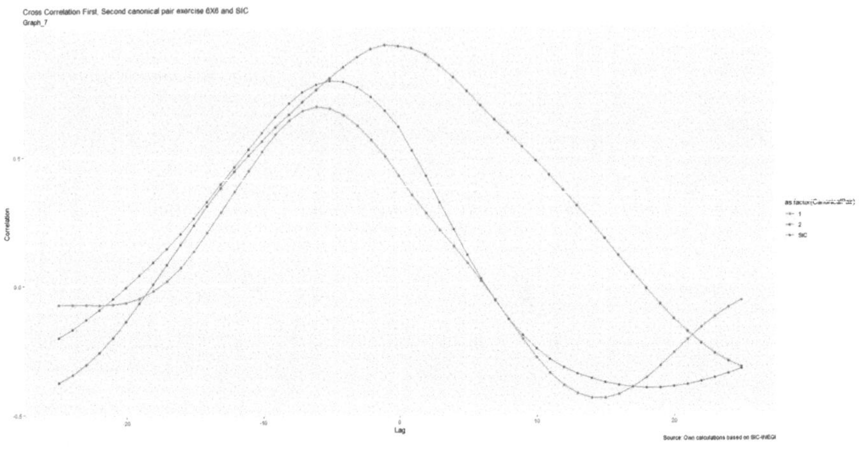

Fig. 8 Cross correlation first 2 canonical pairs and SIC

on cross-correlation functions between each series and IGAE, revealed that some leading or coincident indicators were not performing as anticipated. Consequently, for the coincident indicators case, UrbanUnemp and IncomeGoods+ were identified as potential candidates for removal.

$$SSS_k = \frac{\sum_{t=1}^{T-k} (LI_t - CI_{t+k})^2}{(Var(LI_t) + Var(CI_t))} \tag{5}$$

In this context, LI_t represents the leading indicator, and the coincident indicator is represented by CI_t. Accordingly, we applied our methodology to each coincident indicator, in conjunction with all six leading indicators. This resulted in six 1×6 exercises, as defined by the relevant terminology. As with the previous criterion, UrbanUnemp and IncomeGoods+ did not perform well. As a result, the decision was made to exclude them from the analysis. The coincident series TIIE and TICAM were deemed inadequate based on cross-correlation analysis. After a 4×5 exercise, which excluded the former, and a 4×4 exercise, which excluded both, the decision was made to remove them from further analysis. Table 1 illustrates the values of (5) for these two exercises and the 6×6.

As illustrated in Table 2, the minimum values identified in exercises 4×4, 4×5, and 4×6 are not superior to the optimal solution provided by the SIC. This outcome prompted the authors to formulate a second hypothesis. Comparing the results from the R program with those of a previous version programmed in Fortran revealed minimal discrepancies up to the point of calculating the eigenvalues and vectors of the optimization problem. The discrepancy between the solution vectors was due to the identification conditions utilized by each program. The "eigen" function, programmed in R, requires the first component of the vector to be real, whereas in Fortran, the component that is expected to be real is the one with the highest norm.

Table 1 Series involved in the formation of the SIC-INEGI

Coincident series	
Global Indicator of Economic Activity	IGAE
Industrial Activity Indicator	IndAct
Income Index for the Supply of Retail Goods and Services	IncomeGoods+
Number of Permanently IMSS Insured	IMSS
Urban Unemployment Rate	UURate
Total Imports	Imports
Leading series	
Employment Trend in Manufacturing	ManufEmp
Business Confidence Indicator	BusinessConf
Price and Quote Index of the Mexican Stock Exchange in real terms	MxStockExch
Bilateral Real Exchange Rate México – EUA	MexUSARate
Equilibrium Interbank Interest Rate	TIIE
Standard Poor's 500 Index	SP

Table 2 Series involved in the formation of the SIC-INEGI

Lag	SIC	4 × 4	4 × 5	4 × 6	4 × 4
0	90.81	146.34	191.65	149.57	82.76
1	77.32	125.44	173.95	138.39	61.99
2	66.52	108.94	158.49	129.52	49.22
3	58.57	96.99	145.15	122.76	41.38
4	53.73	89.41	134.15	117.96	39.04
5	52.23	86.00	125.73	114.94	41.52
6	54.00	85.46	119.91	113.5	47.86
7	58.76	87.56	116.64	113.49	57.03
8	66.08	91.32	115.79	114.79	68.01
9	75.45	96.10	117.19	117.36	79.99

They programmed both in R, which resulted in the second 4 × 4 column in Table 2. Notably, none of the previous exercises outperforms the "Fortran" version (We will call "the aligned" version); (Last column in Table 2).

3.2 *Proposal for the System of Coincident and Leading Indicators*

The results indicate that the combination with the lowest forecast error is the aligned 4 × 4 experiment, with coincident indicators IGAE, IndAct, IMSS, and Imports, along with Manufacture, BusinessConf., IncomeGoods+, and SP in the leading group. These will be used as inputs for the calculation of composite indicators. Figure 9 illustrates the squared norms of coherence for the first canonical pair at varying frequencies for the 6 × 6 and the 4 × 4 examples, the removal of information results in suboptimal outcomes. Conversely, the remaining three canonical pairs

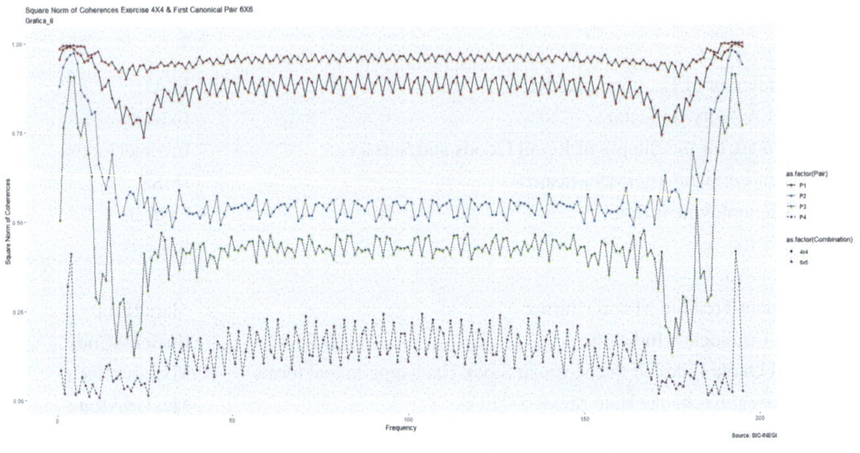

Fig. 9 Squared norms of coherences, four canonical pairs in 4×4, and first canonical pair in 6×6

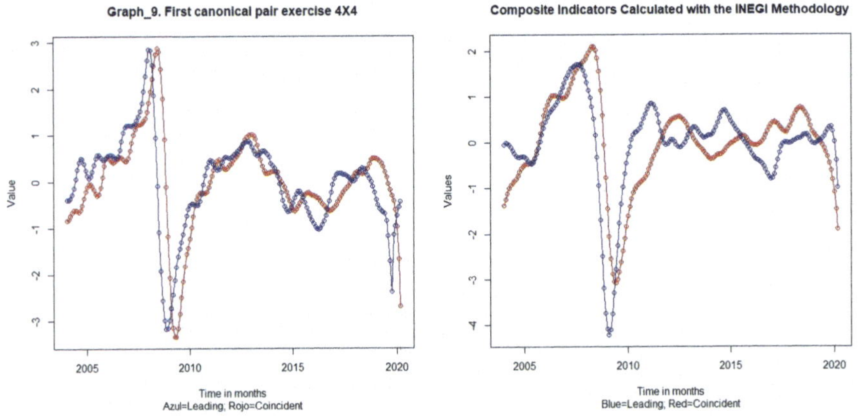

Fig. 10 Left first canonical pair 4×4, Right Composite index INEGI

for the 4×4 exercise do not appear to significantly contribute to the analysis of economic cyclicality, exhibiting behavior that is far from optimal.

Figure 10 illustrates the first canonical pair, which shows a close alignment between the two indicators and a notable positive trajectory from 2008 to 2009. Additionally, we can observe a significant decline in the leading indicator in October 2018, which coincides with a similar decline in the coincident indicator in December of the same year.

The correlation structure for the top two canonical pairs obtained in this exercise, along with that for SIC, is shown in Fig. 11. While the optimal correlation is observed on the SIC side, the difference between it and that of the first canonical pair is insignificant. It is important to note that the first canonical pair has a minimal sum of squares of the forecast error.

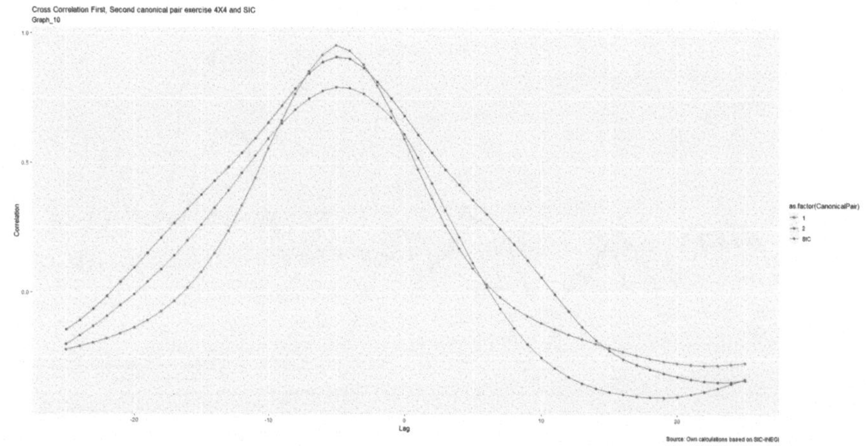

Fig. 11 Cross correlations, 4 × 4 and SIC

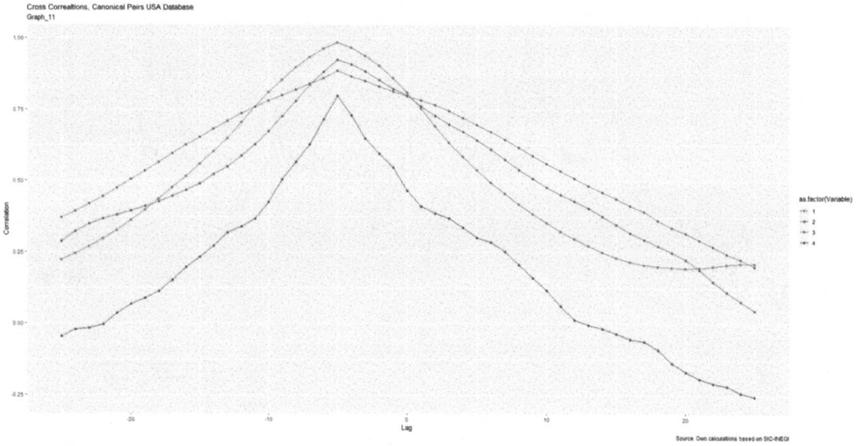

Fig. 12 Cross correlation USA data

3.3 USA Data

To ensure that the results obtained for Mexico were not merely coincidental, we applied the methodology to a dataset comprising a series from the United States (four coincident and seven leading) for the years 2000–2020. The cross-correlation function is depicted in Fig. 12. Notably, the correlation for the first three canonical pairs exceeds 0.8, with a lag of 5 months.

The correlation between the first canonical pair is strong, approaching 1, as illustrated in the accompanying graph (Fig. 13). Similarly, the second canonical pair

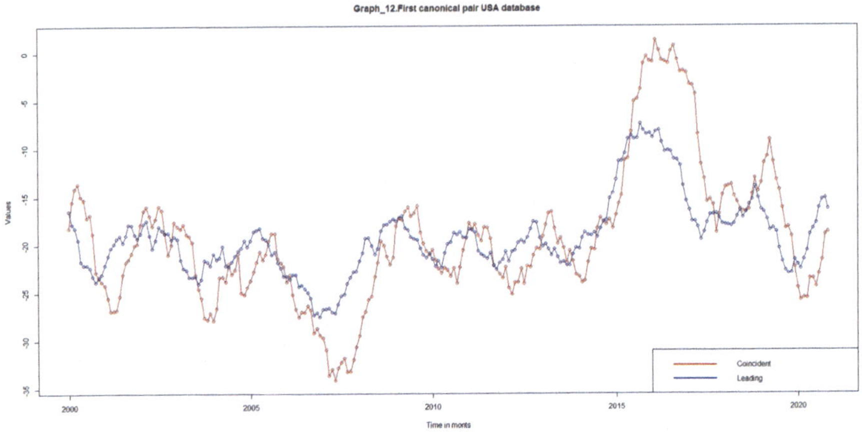

Fig. 13 First canonical pair USA

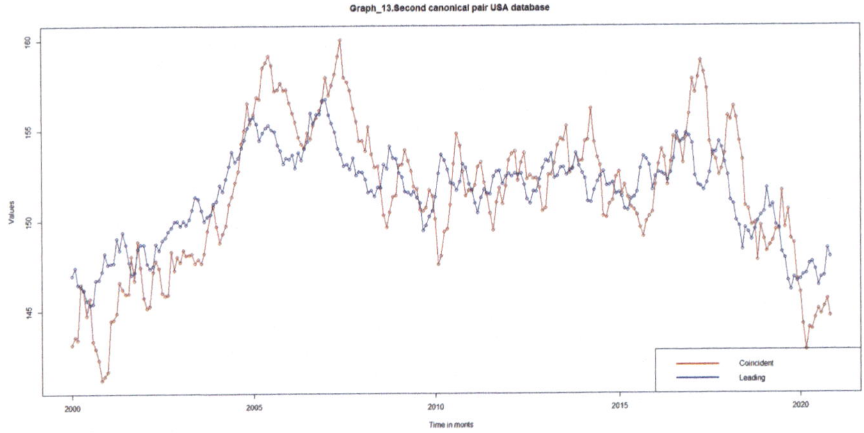

Fig. 14 Second canonical pair

also demonstrates a high level of correlation, as evidenced by a visual inspection (Fig. 14). It is worth noting that the first and second canonical pairs depict distinct aspects of the economic cycle, as indicated by their substantial differences. To gain further insight into their respective influences, we computed cross-correlations between components of the canonical pairs. A1 represents the leading indicator of the first canonical pair, and A2 corresponds to the leading indicator of the second pair, on the other hand, C1 and C2 denote the coincident indicators of the first and second pairs, respectively. Figure 15 illustrates that the canonical pairs are largely independent of each other, validating our assumption.

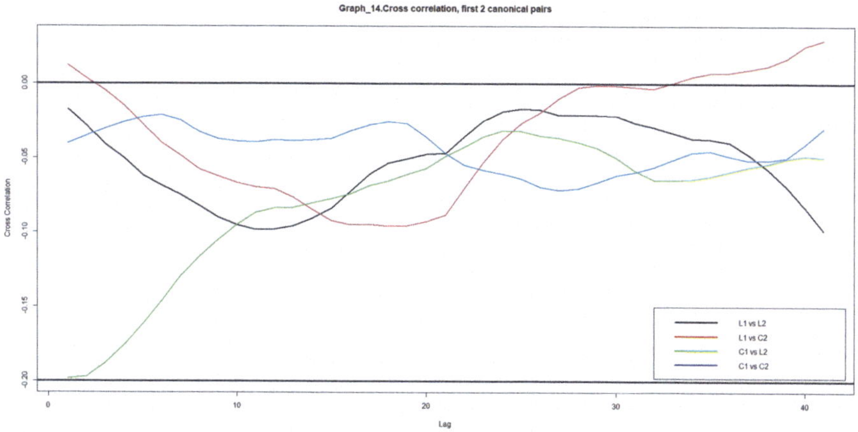

Fig. 15 Cross correlation, first 2 canonical pairs

3.4 Limitations

In the frequency domain, optimal results are achieved, as demonstrated by the first canonical pair of the 6×6 exercise. However, when transitioning back to the time domain, optimal results are not consistently attained for the composite indicators. This is because the solution to the optimization problem is expressed in terms of eigenvectors, which are unique except for multiplication by a complex of norm one. This can be seen in expression (6).

$$\Phi_X(\lambda) = B(\lambda)\Delta(\lambda)\bar{B}^t(\lambda) = B(\lambda)E(\lambda)\Delta(\lambda)\bar{E}^t(\lambda)\bar{B}^t(\lambda) \tag{6}$$

Where the columns of matrix B represent eigenvectors, and the diagonal matrix is formed with the corresponding eigenvalues. Additionally, the arbitrary diagonal matrix, denoted by Δ, has nonzero elements with the form $\delta_{ii} = e^{i\theta_i}$ for each i. This implies that the components of each of the columns of matrix B can be expressed in multiple ways and still represent an optimal solution for the maximization problem. When eigenvectors are repurposed, as in our case, the results can vary significantly. This is demonstrated by Eqs. (7) and (8).

$$\zeta_i(t) = \int_0^{2\pi} \overline{A_J(\lambda)^t} e^{i\lambda t} dZ_X(\lambda) \neq \int_0^{2\pi} \overline{A_J(\lambda)^t} e^{i(\lambda t - \theta_j(\lambda))} dZ_X(\lambda) \tag{7}$$

$$\eta_i(t) = \int_0^{2\pi} \overline{B_J(\lambda)^t} e^{i\lambda t} dY_X(\lambda) \neq \int_0^{2\pi} \overline{B_J(\lambda)^t} e^{i(\lambda t - \theta_j(\lambda))} dZ_Y(\lambda) \tag{8}$$

4 Final Comments and Conclusions

The preceding pages show an example of Brillinger's methodology for deriving composite indicators with maximum correlation in the frequency domain. The optimality achieved in the frequency domain does not necessarily transfer seamlessly to the time domain. This results in suboptimal solutions until the conditions leading to the best solutions are identified. These conditions must be reflected in routines for computing eigenvalues and vectors. Consequently, we are exploring alternative approaches and anticipate obtaining results soon. This methodology is valuable for selecting the most suitable indicators from a group to contribute to the aggregate indicator, considering comparisons beyond mere peaks and valleys with a reference series. Based on our experience, certain series require careful analysis for potential exclusion or reassignment within their group.

References

1. Brillinger, D. R., Time Series, Data Analysis and Theory, Holden Day, Inc., San Fran-cisco, 1981.
2. Bustos, A., Statistical Methodology for leading Indicator, Bulletin of the International Statistical Institute. Contributed papers, Book I. 49th session Florence, Italy, 1993, pp. 189–190
3. Hodrick, R. J. and Prescott, E. C., Postwar U.S. Business Cycles: An Empirical Investi-gation, Journal of Money, Credit and Banking, Vol. 29, No. 1 (Feb. 1997), pp. 1–16, Blackwell Publishing.
4. INEGI, Methodology for the construction of the system of cyclical indicators, National Institute of Statistics and Geography. 2da. ed. Mexico: INEGI, 2015. Reached on 11 Nov., 2020 at http://internet.contenidos.inegi.org.mx/contenidos/productos/prod_serv/contenidos/espanol/bvinegi/productos/nueva_estruc/702825073961.pdf
5. OECD, OECD System of Composite Leading Indicators, OECD, Paris, 2012. Reached on 11 Nov., 2020 at http://www.oecd.org/sdd/leading-indicators/41629509.pdf
6. Pollock, D. S. G. (2016). Econometric filters. Computational Economics, 48, 669–691.
7. Pollock, D. S. G. (2018a). Filters, waves and spectra. Econometrics, 6(3), 1–33.
8. Pollock, D. S. G. (2018b). Trends cycles and seasons: Econometric methods of signal extraction. Econometric Reviews, 37(3), 228–246.
9. Yabuta Yuriko, Third International Seminar on Early Warning and Business Cycle Indicators, New tools for tracking the Mexican Business Cycle, 2010.

A Copula-Based Fully Bayesian Nonparametric Evaluation of Cardiovascular Risk Markers for Normoglycemic Patients in the Mexico City Diabetes Study

Claudia Wehrhahn, Ruth Fuentes-García, Ramsés H. Mena, Fabrizio Leisen, María Elena González-Villalpando, and Clicerio González-Villalpando

Abstract Cardiovascular disease leads the cause of death worldwide, and several studies have been carried out to understand and explore cardiovascular risk markers in normoglycemic and diabetic populations. In this work, we explore the association structure between hyperglycemic markers and cardiovascular risk markers controlled by triglycerides, body mass index, age, and sex, for the normoglycemic population in The Mexico City Diabetes Study. Understanding the association structure contributes to the assessment of additional cardiovascular risk markers in this low-income urban population with a high prevalence of classic cardiovascular risk biomarkers. The association structure is measured by conditional Kendall's tau, defined by means of conditional copula functions. The latter are in turn modeled under a fully Bayesian nonparametric approach, which allows the complete shape

C. Wehrhahn (✉)
HDI Insurance, Santiago, Chile

R. Fuentes-García
Faculty of Sciences, Universidad Nacional Autónoma de México, Mexico City, Mexico
e-mail: rfuentes@ciencias.unam.mx

R. H. Mena
IIMAS, Universidad Nacional Autónoma de México, Mexico City, Mexico
e-mail: ramses@sigma.iimas.unam.mx

F. Leisen
Department of Mathematics, King's College, London, UK
e-mail: fabrizio.Leisen@kcl.ac.uk

M. E. González-Villalpando
Center for Studies in Diabetes A.C., Mexico City, Mexico

National Institute of Public Health, Mexico City, Morelos, Mexico

C. González-Villalpando
Center for Studies in Diabetes A.C., Mexico City, Mexico

National Institute of Public Health, Mexico City, Morelos, Mexico

J. A. Christen et al. (eds.), *Statistics, Society and Environment*, Springer Proceedings in Mathematics & Statistics 479, https://doi.org/10.1007/978-3-031-78401-9_10

of the copula function to vary for different values of controlled covariates, which mediate the association.

Keywords Cardiovascular risk marker · Carotid intima-media thickness · Conditional copula function · Dependent stick-breaking process · Hyperglycemic risk marker

1 Introduction

Considerable research has been carried out to understand the complex pathogenesis of Type 2 Diabetes (T2D) as well as its role as a cardiovascular risk factor in Mexico [8, 14]. Although a higher risk of cardiovascular disease is expected for diabetic patients, this relation is not so clear for normoglycemic patients, i.e., patients with normal sugar blood levels. In normoglycemic patients, the identification and understanding of possible cardiovascular risk markers are important to monitor the potential development of cardiovascular disease [32]. One question that has arisen in the literature is whether the glycated hemoglobin (HbA1c) test can provide more information about cardiovascular risk compared to that provided by the 2-hour postchallenge glucose (PG2-H) test. Huang et al. [17] evaluated the association between hyperglycemic and cardiovascular risk markers, which has been reported as non consistent for non-diabetic subjects, in a population from Shanghai. In particular, using linear regression models considering the quartiles of the glycemic markers and adjusting by sex, smoking and drinking status, body mass index (BMI), blood pressure, and serum lipids, they analyzed the association between HbA1c and PG2-H with the carotid intima-media thickness (CIMT) test, a cardiovascular risk marker. They reported that the population of normoglycemic subjects they analyzed had a higher association between HbA1c and CIMT than between PG2-H and CIMT, suggesting that HbA1c could be more informative of cardiovascular risk. To contribute to this discussion, our analysis aims to provide guidelines of possible association between cardiovascular and hyperglycemic risk markers, for normoglycemic patients in the Mexico City Diabetes Study (MCDS). The MCDS is a prospective, population-based, and clinico-epidemiological study about T2D and follows a low income cohort of elder urban participants in Mexico City. We focus on the normoglycemic patients from the MCDS, and our aim is to investigate the use of hyperglycemic markers such as PG2-H or HbA1c as cardiovascular risk markers, controlled by relevant covariates. Although our interest is not on causal relations, covariates such as triglycerides, BMI, age, and sex could mediate the association; therefore, they are included in our study.

We study the association structure between blood sugar measurements and cardiovascular risk markers by means of the Kendall rank correlation coefficient, hereafter named Kendall's tau, which measures the ordinal association between variables. Specifically, we model Kendall's tau between PG2-H and CIMT and between HbA1c and CIMT for different values of covariates such as triglycerides levels, BMI, age, and sex. The use of classical nonparametric rank correlation tests in the analysis of the MCDS dataset results in non-statistical significance; moreover, it does not highlight the dynamics of the association structure for different sets of covariates. An alternative direction is to model conditional copula functions, which in turn define conditional Kendall's tau. This strategy provides a complete characterization of the conditional association structure between cardiovascular and hyperglicemic markers, modulated by covariates. However, modeling conditional copula functions under a parametric approach leads to a model selection problem. We explore a more flexible route that avoids this selection and provides a better understanding of the association structures. Therefore, we undertake a Bayesian nonparametric method to model the covariate dependent copula functions between PG2-H and CIMT and between HbA1c and CIMT for varying values of covariates. Through a post processing analysis, we obtain the dynamics of the corresponding conditional Kendall's tau.

Bayesian nonparametric (BNP) models have been widely used, and their advantages, compared to their parametric counterparts, are abound. These models relax standard parametric assumptions, such as the choice of a particular family of distributions or the number of components that a mixture model requires. Due to their advantages, the use of BNP models for estimating copula functions has grown significantly in recent years. Regarding unconditional copula functions [35] and [36] proposed BNP models for copula density estimation based on infinite mixtures of Gaussian and skew-normal copulas, respectively. Burda and Prokhorov [4] modeled the dependency structure between random variables by first modeling the marginals as infinite mixtures, and then linking them through a random Bernstein polynomial copula function, while [26] modeled the copula function using Dirichlet polya trees. In the context of predictor-dependent copula function modeling [16, 22], and [21] considered transformations of Gaussian processes in the definition of the copula function.While these methods are attractive, they necessitate the choice of a particular copula function. Dalla et al. [20] extended the model proposed by Wu et al. [36] and defined an infinite mixture of conditional Gaussian copula functions, which is flexible, but relies on the selection of a calibration function that relates the correlation structure of the copula function with the predictors. Our proposal extends the model proposed by Dalla et al. [20] and overcomes the aforementioned drawbacks.

Here, we model conditional copula functions using a BNP strategy. Under this modeling framework, a prior distribution is placed on sets of predictor-dependent random measures involved in the definition of the model. More specifically, our fully BNP approach considers a predictor-dependent stick-breaking prior distribution for the collection of predictor-dependent random measure, as proposed by MacEachern [23]. In this framework, for each vector of covariates, the conditional copula function is represented as an infinite mixture of copula functions. This allows copula functions to vary flexibly across covariates, resulting in a dynamic and adaptable characterization of the association structure for different covariate values. As will be shown later, this modeling approach, through the borrowing of information between covariates and its inherent flexibility, offers improvements over the framework proposed by Dalla et al. [20]. BNP models based on the dependent stick-breaking and other extensions have been proposed by De Iorio et al. [6], Gelfand et al. [9], De Iorio et al. [5], Rodriguez and Dunson [29], Barrientos et al. [3], among others.

The rest of this chapter is organized as follows. Section 2 describes the MCDS and motivates the analysis for the real dataset while Sect. 3 provides background and the description of the model. Section 4 illustrates the performance of the model in synthetic datasets while Sect. 5 describes the association analysis for the MCDS. Finally, Sect. 6 concludes the manuscript.

2 The Mexico City Diabetes Study

The MCDS [13] study began in 1989 with the identification of a homogeneous low-income site. All men and non-pregnant women aged 35–64 years were defined as eligible. The research protocol, informed consent, procedures, and methods were approved by the Institutional Review Board of the Center for Studies in Diabetes and all participants gave informed consent.

At baseline (1989–1990), a total of 3,319 participants were interviewed and 2,282 examined (from a total of 3,505 eligible individuals). The final cohort (interviewed and examined) was composed of 941 men and 1,341 women. During the last phase (2008–2009), for a total of 1,174 participants, in addition to the standardized exam and oral glucose tolerance test, an HbA1c measurement was included. T2D was diagnosed using American Diabetes Association criteria: fasting plasma glucose concentration 126 mg/dL and/or a 2-hour plasma glucose concentration 200 mg/dL after a standard 75-g glucose load. Participants who self-reported a history of diabetes and were taking oral glucose-lowering agents were considered to have T2D, regardless of their plasma glucose values. Pre-diabetes (PD) was diagnosed when an individual had a fasting plasma glucose of 100–125 mg/dL and/or a 2-hour postglucose load between 140 and 199 mg/dL.

In our analysis, we consider the subset of normoglycemic individuals from the MCDS at the last phase of the study who had the CIMT measured in phase 3 and who had measurements for all the considered covariates. Since the CIMT is prone to measurement error, values lower than the first quartile and greater than the third quartile were excluded. The resulting datasets contain $n = 246$ individuals. As covariates we include triglycerides, BMI, sex, and age, the latter discretized into four categories, considering reference values. Table 1 provides summaries for the variables involved in the analysis. The median age of the participants is 59 years, and 58.9% are women. Although our analysis focuses on the normoglycemic subset from the MCDS, we note that more than 84% of this group shows overweight or obesity ($BMI > 25$) and more than 42% has borderline, high, or very high triglycerides levels ($Triglycerides > 150$).

To contribute in the assessment of potential use of HbA1c as a cardiovascular risk marker, we explore the association between HbA1c and CIMT, controlled by triglycerides, BMI, age, and sex, and compare these results with those obtained

Table 1 Descriptive statistics for variables involved in the analysis. Summaries for continuous variables are presented as median (interquartile range) while discrete variables are presented as count (%)

Continuous marker		Continuous covariate		Discrete covariate	
HbA1c	6.3 (1.1)	Triglycerides	141.5 (91)	Female sex	145 (58.9 %)
PG2-H	102.5 (30)	BMI	28.3 (4.98)	$55 \leq Age < 60$	78 (31.71 %)
CIMT	0.64 (0.15)			$60 \leq Age < 65$	49 (19.92 %)
				$Age \geq 65$	66 (26.83 %)

from analyzing the association between PG2-H and CIMT, controlled by the same set of covariates. Our interest is not on causal relations between HbA1c and CIMT or PG2-H and CIMT. As a matter of fact, if we were to fit a linear regression model for CIMT with HbA1c and PG2-H as independent variables, none of them would be statistically significant, and the inclusion of additional covariates would not change these results.

3 Background and Methods

It is well known that copula functions, introduced in the statistical literature by Sklar [31], describe the dependency structure between two or more random variables. Roughly speaking, copulas link the univariate marginal distribution function of random variables with their joint cumulative distribution function (c.d.f.). Since our interest is to model the dependency structure between pairs of random variables, in what follows we give a brief description of two-dimensional copulas.

A two-dimensional copula function, denoted C, is a c.d.f. on $[0, 1]^2$ such that $C(u, 0) = C(0, u) = 0$, $C(u, 1) = C(1, u) = u$, for every $u \in [0, 1]$, and $C(A(u_1, u_2)) \geq 0$ for every set $A(u_1, u_2) \subseteq [0, 1]^2$. According to Sklar's theorem, the joint c.d.f., denoted H, of two random variables, say Y_1 and Y_2, with univariate marginals F_1 and F_2, respectively, can be linked through a copula function, which is unique, when F_1 and F_2 are continuous. The relation is as follows: $H(y_1, y_2) = C(F_1(y_1), F_2(y_2))$. When the bivariate c.d.f. has a probability density function (p.d.f.), denoted h, it holds that $h(y_1, y_2) = c(F_1(y_1), F_2(y_2)) f_{Y_1}(y_1) f_{Y_2}(y_2)$, where

$$c(v_1, v_2) = \frac{\partial C(v_1, v_2)}{\partial v_1 \partial v_2},$$

is the copula density and f_{Y_1} and f_{Y_2} are the p.d.f.s of Y_1 and Y_2, repectively. A formal definition of copula functions, their properties, and more, can be found in [25].

Given the practicability for covariate inclusion, here we concentrate on the family of Gaussian copulas. Let $\Phi(y_1, y_2 \mid \rho)$ denote the c.d.f. of a bivariate normal distribution with zero mean vector and correlation parameter ρ and let $\Phi^{-1}(p)$ denote the inverse c.d.f. of a standard normal distribution, $p \in (0, 1)$. The Gaussian copula function, denoted $C^G(u_1, u_2 \mid \rho)$, of the bivariate vector $(u_1, u_2) \in [0, 1]^2$, with correlation parameter $\rho \in (-1, 1)$, is given by $C^G(u_1, u_2 \mid \rho) := \Phi(\Phi^{-1}(u_1), \Phi^{-1}(u_2) \mid \rho)$. The associated Gaussian copula density, denoted $c^G(u_1, u_2 \mid \rho)$, is given by

$$c^G(u_1, u_2 \mid \rho) = \\ |\Sigma_\rho|^{-1/2} \exp\left\{-\frac{1}{2}(\Phi^{-1}(u_1), \Phi^{-1}(u_2))(\Sigma_\rho^{-1} - \boldsymbol{I}_2)\begin{pmatrix}\Phi^{-1}(u_1)\\ \Phi^{-1}(u_2)\end{pmatrix}\right\}, \tag{1}$$

where $\boldsymbol{I}_2$ is the two-dimensional identity matrix and

$$\Sigma_\rho = \begin{pmatrix} 1 & \rho \\ \rho & 1 \end{pmatrix}.$$

Copula functions play a key role in the study of the association between random variables. In particular, one scale-invariant measure of association is Kendall's tau, denoted τ. Specifically, Kendall's tau measures the concordance between random variables, is defined by means of copulas, and is given by

$$\tau = 4E\left[C(U_1, U_2)\right] - 1 = 4\int_0^1\int_0^1 C(u_1, u_2)dC(u_1, u_2) - 1.$$

Our interest is to model the dependence structure between cardiovascular and hyperglycemic risk markers, controlled by relevant variables. Thus, we need the dependence structure to be dynamic for varying values of such variables. There is a rich literature on conditional copula functions and measures of association. See [1, 2, 11, 12, 15, 19, 27, 33], among others.

Conditional joint distributions are completely described by the conditional marginals and a conditional copula function, which is unique when the conditional marginals are continuous. According to Sklar's theorem, the relation is as follows: $H(y_1, y_2 \mid \boldsymbol{x}) = C(F_1(y_1 \mid \boldsymbol{x}), F_2(y_2 \mid \boldsymbol{x}) \mid \boldsymbol{x})$, where $C(\cdot \mid \boldsymbol{x})$ is the predictor-dependent copula function that describes the dependence structure between two random variables for varying values of the predictor $\boldsymbol{x}$. This, in turn, directly defines predictor dependent measures of association such as conditional Kendall's tau, given by

$$\tau(\boldsymbol{x}) = 4E\left[C(U_1, U_2 \mid \boldsymbol{x})\right] - 1 = 4\int_0^1\int_0^1 C(u_1, u_2 \mid \boldsymbol{x})dC(u_1, u_2 \mid \boldsymbol{x}) - 1. \tag{2}$$

As stated before, the purpose here is to robustly model predictor-dependent copula functions to describe the predictor dependent association structure between cardiovascular and hyperglycemic risk markers. Dependent BNP models provide a flexible tool for such data modeling framework. In what follows, we describe our covariate adjusted copula function and its use to measure the conditional association structure between cardiovascular and hyperglycemic risk markers.

We consider a Dirichlet process mixture of bivariate Gaussian copula densities model with predictor-dependent mixing distribution. Specifically, we use a dependent stick-breaking prior distribution for the set of predictor dependent random measures, as proposed by MacEachern [23], who extended the well-known stick-breaking process prior proposed by Sethuraman [30] to accommodate for covariate dependency. The model can be written as

$$c(u_1, u_2 \mid \boldsymbol{x}) = \int c^G(u_1, u_2 \mid z) G_{\boldsymbol{x}}(dz), \qquad (u_1, u_2) \in [0, 1]^2, \tag{3}$$

where c^G is defined as in Eq. (1) and the set $\{G_{\boldsymbol{x}} : \boldsymbol{x} \in \mathscr{X}\}$ follows a dependent stick-breaking process prior such that, for each $\boldsymbol{x} \in \mathscr{X}$, the predictor-dependent random measure is given by $G_{\boldsymbol{x}}(\cdot) = \sum_{j=1}^{\infty} w_j(\boldsymbol{x}) \delta_{\rho_j(\boldsymbol{x})}(\cdot)$, with $w_j(\boldsymbol{x}) = v_j(\boldsymbol{x}) \prod_{l<j} [1 - v_l(\boldsymbol{x})]$, where the stick-breaking variables, $\{v_j(\boldsymbol{x})\}_{j=1}^{\infty}$, and correlations, $\{\rho_j(\boldsymbol{x})\}_{j=1}^{\infty}$, are defined by transformations of real-valued stochastic processes to be specified further. Indeed, notice that both, weights and correlations, are dynamic. To define $v_j(\boldsymbol{x})$ and $\rho_j(\boldsymbol{x})$, we exploit the relation between Gaussian processes and linear regression models and consider suitable transformations. Note that the model in Eq. (3) can be equivalently written as

$$c(u_1, u_2 \mid \boldsymbol{x}) = \sum_{j=1}^{\infty} w_j(\boldsymbol{x})\, c^G(u_1, u_2 \mid \rho_j(\boldsymbol{x})), \qquad (u_1, u_2) \in [0, 1]^2. \tag{4}$$

Under this approach, the copula density is flexibly modeled through a mixture of bivariate Gaussian copula densities that vary and accommodate for different values of the predictor. We consider $v_j(\boldsymbol{x})$ defined by the logistic transformation of the linear predictor, i.e., $v_j(\boldsymbol{x}) = e^{\boldsymbol{x}^t \boldsymbol{\beta}_j^v}/(1 + e^{\boldsymbol{x}^t \boldsymbol{\beta}_j^v})$, with $\boldsymbol{\beta}_j^v \overset{iid}{\sim} N_p(\boldsymbol{\mu}^v, \Sigma^v)$, $j \geq 1$. Similarly, the $\rho_j(\boldsymbol{x})$ processes are defined by $\rho_j(\boldsymbol{x}) = 2/(|\boldsymbol{x}^t \boldsymbol{\beta}_j^\rho| + 1) - 1$, with $\boldsymbol{\beta}_j^\rho \overset{iid}{\sim} N_p(\boldsymbol{\mu}^\rho, \Sigma^\rho)$, where $N_p(\boldsymbol{\mu}, \Sigma)$ denotes a p-dimensional normal distribution with parameters $\boldsymbol{\mu} \in \mathbb{R}^p$ and Σ a $(p \times p)$-dimensional positive definite matrix. We termed this model as dependent Dirichlet process mixture of copula densities (DDPMC) model. A sufficient and necessary condition for the weights to add up to one in Eq. (4) is that for every $\boldsymbol{x} \in \mathscr{X}$, $\sum_{j=1}^{\infty} \log\left[1 - E(v_j(\boldsymbol{x}))\right] = -\infty$ (Lemma 1 in [18]).

The conditional association structure between cardiovascular and hyperglycemic risk markers can be described by the covariate-dependent Kendall's tau, which as defined in Eq. (2), depends on the dynamic copula functions. Under the DDPMC model, inference for the association between variables of interest is easily achieved for varying values of the predictors.

4 Model Validation with Simulated Data

We illustrate the performance of the model in two simulation scenarios that, to some extent, resemble the observed features in the analysis of the MCDS data set. For each simulation scenario, one sample of size 250 was generated from the bivariate joint distribution defined by a copula function and standard normal marginals. In order to have data points that closely approximate samples U_{i1} and U_{i2} from the copula densities, we consider the pseudo observations $\tilde{U}_{i1} = R_{i1}/(n+1)$ and $\tilde{U}_{i2} = R_{i2}/(n+1)$, where R_{il} denote the rank of Y_{il} among $Y_{1l}, \ldots, Y_{nl}$, $l = 1, 2$, and (Y_{i1}, Y_{i2}) denote the pairs generated from the joint distribution, as discussed by Genest et al. [10].

Under both scenarios, uniform on (0, 1) values were generated for the predictor. The first simulation scenario, named Scenario I, is defined by a predictor-dependent elliptical t copula with correlation parameter $\rho(x) = x$. Kendall's tau for Scenario I is an increasing positive function of the predictor that varies from 0.006 to 0.910 for values of the predictor between 0.010 and 0.990. The second simulation scenario, named Scenario II, is defined by a mixture of two predictor dependent copula functions: a t copula with correlation parameter $\rho(x) = -x(1-x)^2$ and a Gumbel copula with parameter $\alpha(x) = x^2(1-x) + 1$, and predictor-dependent weights. Under this scenario, Kendall's tau is an almost constant function of the predictor that varies between 0.0 and 0.015 showing a very low positive association between the random variables.

We compare the performance of the proposed methodology with the model proposed by Dalla et al. [20], which corresponds to a Dirichlet process mixture of conditional Gaussian copula densities (DPMCG) model. Under this setting, a stick-breaking process prior is assumed for a single random measure, which is fundamentally different from our proposal in which a dependent stick-breaking process prior is assumed for the collection of covariate-dependent random measures, allowing them to flexible vary for different values of the predictor. Additionally, under the DPMCG model, the dependency on covariates is driven by the calibration function, denoted $\theta(\boldsymbol{x} \mid \boldsymbol{\beta})$, $\boldsymbol{\beta} \in \mathbb{R}^p$, for which a specific choice is required. The DPMCG model is given by

$$c(u_1, u_2 \mid \boldsymbol{x}) = \sum_{j=1}^{\infty} w_j \, c^G(u_1, u_2 \mid \rho_j(\boldsymbol{x})),$$

where c^G is defined as in Eq. (1), $\rho_j(\boldsymbol{x}) = 2/(|\theta(\boldsymbol{x} \mid \boldsymbol{\beta})| + 1) - 1$, and $w_j = v_j \prod_{l<j}(1 - v_l)$, where $\boldsymbol{\beta} \overset{iid}{\sim} N_p(\boldsymbol{\mu}, \Sigma)$, $v_j \overset{iid}{\sim} Beta(1, \alpha)$, and $\alpha > 0$. Regarding the calibration function we use $\theta(\boldsymbol{x} \mid \boldsymbol{\beta}) = \beta_1 + \beta_2 x^2$, which results as the optimal choice of those considered in [20].

We compare the performance of both models by means of an estimate to the integrated L_1 distance, denoted $\widehat{IL}_1$, between the true conditional Kendall's tau function, denoted $\tau^{true}(x)$, and its conditional estimate, denoted $\widehat{\tau}(x)$, given by

$$\widehat{IL}_1 = \frac{1}{L}\sum_{l=1}^{L}\left|\widehat{\tau}(x_l) - \tau^{true}(x_l)\right|, \tag{5}$$

where $\widehat{\tau}$ is given by the pointwise posterior predictive median and $\{x_l\}_{l=1}^{L}$ denotes an equally spaced grid of values on (0, 1).

Under both simulated scenarios, the specification of the DDPMC model is completed by considering a truncation level of N for the random measure. Indeed, as we will describe further, we rely on the `nimble` package [7], which uses truncation of sum terms in the species sampling representation of the random probability measure, as an algorithmic shortcut.

Since only one covariate is simulated, $p = 2$, and the hyperparameters were set to $\boldsymbol{\mu}^{v} = \boldsymbol{\mu}^{\rho} = (0, 0)^t$ and $\boldsymbol{\Sigma}^{v} = \boldsymbol{\Sigma}^{\rho} = 2.25\boldsymbol{I}_2$, where $\boldsymbol{I}_p$ denotes the $(p \times p)$ identity matrix. Under this specification, the prior distribution induced for the stick-breaking variables assigns positive probability to the (0, 1) interval and the prior distribution induced for the correlation parameter is concentrated on the $(-0.70, 0.99)$ interval, for varying values of the predictor. Note that this prior distribution for the correlation parameter is adequate for both simulation scenarios. Following a similar model specification, for the DPMCG model, we consider $\boldsymbol{\mu} = (0, 0)^t$ and $\boldsymbol{\Sigma} = 2.25\boldsymbol{I}_2$, which also results in a prior distribution for the correlation parameter that is concentrated on the $(-0.70, 0.99)$ interval, and $\alpha = 1$.

The Markov chain Monte Carlo (MCMC) posterior sampling scheme for the proposed DDPMC model sequentially updates the coefficients of the linear predictors defining v_j and ρ_j with a multivariate slice sampler step [24].

We use the recently developed BNP functionality of the `nimble` package [7, 34] from the R statistical software to generate posterior samples of the DPMCG model. The sampling scheme is based on a truncation to a level N of the random measure and on random walk Metropolis–Hastings and categorical samplers for the parameters in the model.

Under both models, a single Markov chain was generated for each simulated dataset. For the proposed DDPMC model, a Markov chain of size 110,000 was generated, and inference is based on a reduced chain of size 10,000 obtained by saving one every 10 iterations, after a burn-in period of 10,000, for both datasets. For the DPMCG model, Markov chains of size 1,510,000 and 210,000 were generated for Scenarios I and II, respectively. For both scenarios, inferences are based on a reduced chain of size 10,000 obtained after a burn-in period of 10,000 and thinning every 150 and 20 iterations for Scenarios I and II, respectively. Convergence of the posterior sample was evaluated with standard convergence tests as implemented in the CODA R library [28] and by examining trace plots.

Figure 1 displays the covariate-dependent Kendall's tau estimate (pointwise posterior predictive median), its corresponding 95% credible bands, and the true covariate-dependent Kendall's tau for Scenario I and Scenario II and under both models. Note that under Scenario I, the proposed DDPMC model is able to recover the dynamic association structure between the random variables for every value of

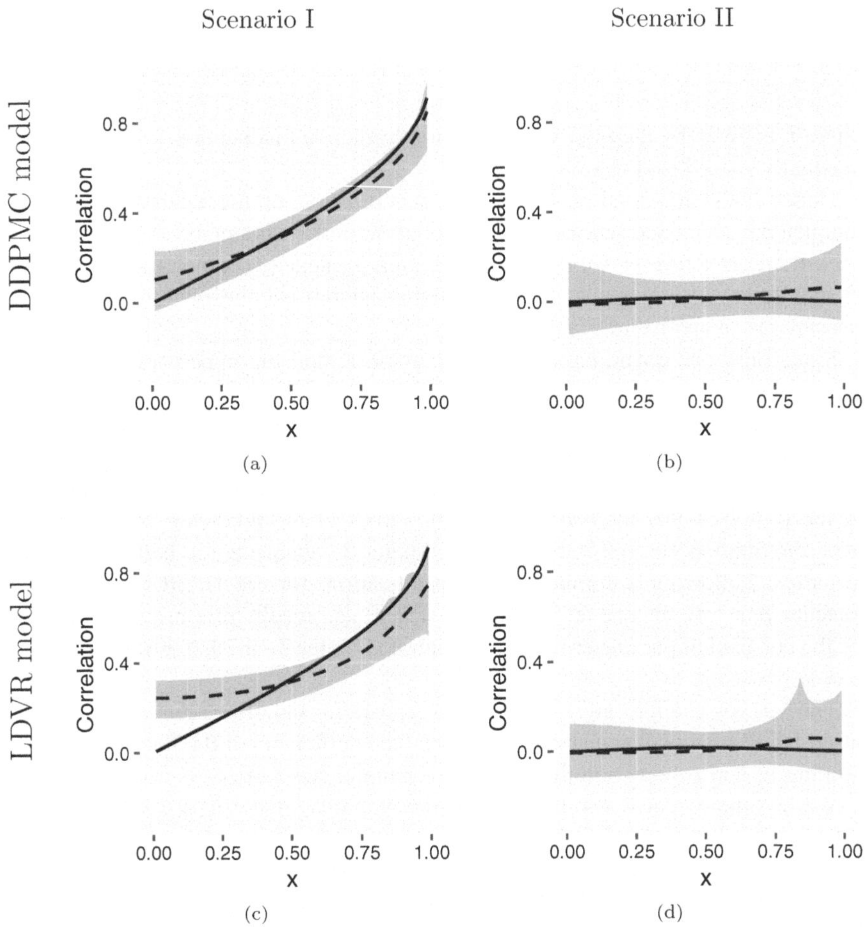

Fig. 1 Simulated Data: Kendall's tau estimate (dashed line) and true Kendall's tau (solid line) as a function of the predictor, x, for Scenario I (first column) and Scenario II (second column). Results are displayed for the DDPMC model (first row) and DPMCG model (second row). Gray areas represent a 95% credible band

the predictor, while the DPMCG model is not able to recover the true association structure over all possible values of the covariate. The DPMCG model provides good estimates for values greater than 0.3 of the covariate, but struggles for smaller values of the covariate. Under the proposed DDPMC model, the true values of Kendall's tau lie within the 95% credible band for every value of the predictor, which is not the case for the DPMCG model. Under Scenario II, Kendall's tau is estimated as an almost constant function of the predictor, with very low values. The true Kendall's tau lies within the 95% credible band for every value of the predictor, under both modeling settings.

In order to quantify the differences of the results using both methodologies, we compute estimates to the integrate L_1 distances, given by Eq. (5). For simulation Scenario I, these values are 0.0387 and 0.0934, under the proposed DDPMC model and the DPMCG model, respectively. Correspondingly, for Scenario II, these values are 0.0210 and 0.0214. These results are consistent with the previous discussion.

5 Application

Let Y_{i1} and Y_{i2} be measurements of cardiovascular and hyperglycemic risk markers of a normoglycemic patient i from the MCDS, respectively. Here, measurements of CIMT represent the cardiovascular risk marker, and measurements of HbA1c and PG2-H represent the hyperglycemic markers. In order to determine which hyperglycemic marker shows a stronger relationship with cardiovascular risk, we model the dependence structure between each one and the CIMT. Additionally, we consider that such dependence structure is affected by the triglycerides level, BMI, age, and sex of patient i, encoded in a predictor vector $\boldsymbol{x}_i$.

For both real datasets, the specification of the model is completed by considering a truncation level of N for the random measure. Since the predictor age was discretized into four categories, the real datasets contain $p = 7$ predictors. Here, the mean hyperparameters were set to $\boldsymbol{\mu}^{\upsilon} = \boldsymbol{\mu}^{\rho} = \mathbf{0}^t$ and the covariances Σ^{υ} and Σ^{ρ} were defined by block diagonal matrices of the form

$$\Sigma^{\upsilon} = \begin{pmatrix} c_1^{\upsilon}(\tilde{\mathbb{X}}^t\tilde{\mathbb{X}})^{-1} & \mathbf{0} \\ \mathbf{0} & c_2^{\upsilon}(\overline{\mathbb{X}}^t\overline{\mathbb{X}})^{-1} \end{pmatrix} \quad \text{and} \quad \Sigma^{\rho} = \begin{pmatrix} c_1^{\rho}(\tilde{\mathbb{X}}^t\tilde{\mathbb{X}})^{-1} & \mathbf{0} \\ \mathbf{0} & c_2^{\rho}(\overline{\mathbb{X}}^t\overline{\mathbb{X}})^{-1} \end{pmatrix},$$

where $\tilde{\mathbb{X}}$ and $\overline{\mathbb{X}}$ denote the design matrices based on the continuous and discrete covariates, respectively [37]. The scaling constants c_j^{υ} and c_j^{ρ}, $j = 1, 2$, control the linear predictor and were selected so that the stick-breaking variables range between 0.02 and 0.99 and the correlations range between −0.70 and 1.00, for every value of the predictor, respectively.

As in the simulation scenario, the MCMC posterior sampling scheme sequentially updates the coefficients of the linear predictors defining υ_j and ρ_j with a multivariate slice sampler step. A single Markov chain of size 300,000 was generated for each real dataset, and inferences are based on a reduced chain of size 5,000 obtained after discarding the first 200,000 iterations and saving one every 20 iterations. Convergence of the posterior sample was evaluated with standard convergence tests as implemented in the CODA R library and by examining trace plots.

Figures 2, 3, and 4 display conditional Kendall's tau between CIMT and PG2-H (dark gray) and between CIMT and HbA1c (light gray) for the normoglycemic participants in the MCDS for three different levels of BMI, namely 22, 27 and 32, respectively. We consider the values 22, 27, and 32 of BMI as indicative of

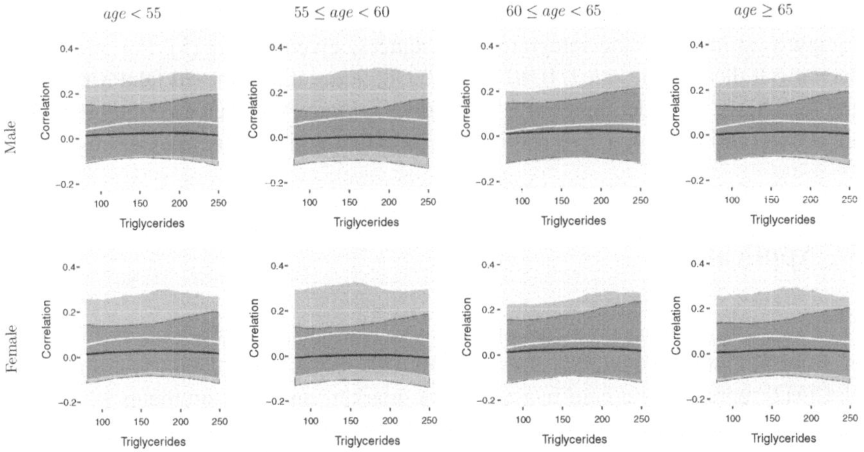

Fig. 2 Mexico City Diabetes Study Data: section plot of conditional Kendall's tau between CIMT and PG2-H (dark gray) and between CIMT and Hb1Ac (light gray) when BMI equals 22. Panels display the estimate of Kendall's tau for varying values of triglycerides, sex, and age category for normoglycemic participants: the continuous line is the estimate (dark gray for Kendall's tau between CIMT and PG2-H and light gray for Kendall's tau between CIMT and Hb1Ac) while the gray region is a 90% credible band. The y-axis shows values for Kendall's tau while x-axis shows values of triglycerides

individuals with normal weight (defined as BMI between 18.5–24.9), overweight (defined as BMI between 25–29.9), and obese (BMI 30 or above), respectively. Panels display conditional Kendall's tau estimate (pointwise posterior predictive median) and 90% credible bands for triglycerides in a range between $(80, 250)$, males, females, and the four categories of age. From these figures, it is clear that the association structure, described by Kendall's tau, varies for different value of the covariates (trygliceride s, sex, and age) in shape and strength. In the three figures, the association between PG2-H and CIMT for the normoglycemic patients is a slightly decreasing positive function of the triglycerides levels for both, male and female, through all age categories. Although the association is relatively small, it is stronger for higher levels of BMI. The association structure between HbA1c and CIMT shows to be a roughly constant function of triglycerides for both male and female through all age categories, when BMI equals 22, whereas these associations become increasing functions of the triglycerides as BMI increases. Although appealing, the higher values of Kendall's tau between HbA1c and CIMT obtained for higher values of triglycerides do not seem to differ greatly from the corresponding Kendall's tau between PG2-H and CIMT.

In order to gain better understanding regarding which hyperglycemic marker would provide more evidence of association with the cardiovascular risk marker for varying values of the predictors, we compute the test statistic of the nonparametric Kendall's tau correlation test. For a large sample size, the predictor-dependent test statistic $S(\tau(\boldsymbol{x})) = \tau(\boldsymbol{x})/\sqrt{\frac{2(2n+5)}{9n(n-1)}}$, where n is the sample size (here $n = 246$),

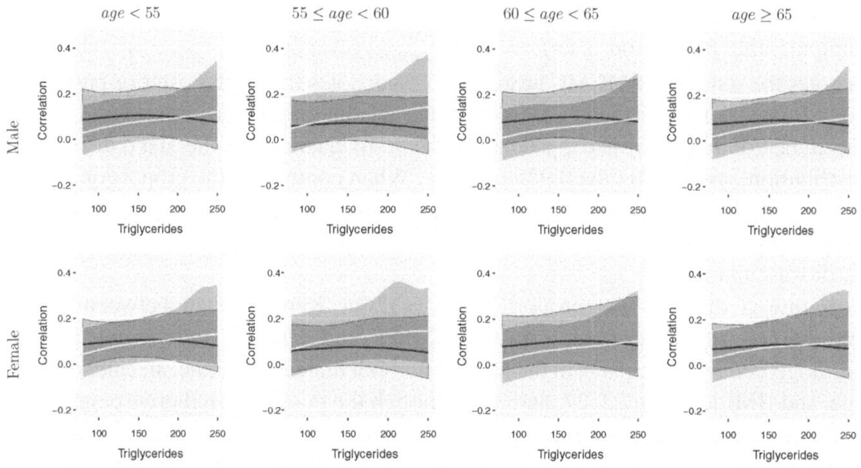

Fig. 3 Mexico City Diabetes Study Data: section plot of conditional Kendall's tau between CIMT and PG2-H (dark gray) and between CIMT and Hb1Ac (light gray) when BMI equals 27. Panels display the estimate of Kendall's tau for varying values of triglycerides, sex, and age category for normoglycemic participants: the continuous line is the estimate (dark gray for Kendall's tau between CIMT and PG2-H and light gray for Kendall's tau between CIMT and Hb1Ac) while the gray region is a 90% credible band. The y-axis shows values for Kendall's tau while x-axis shows values of triglycerides

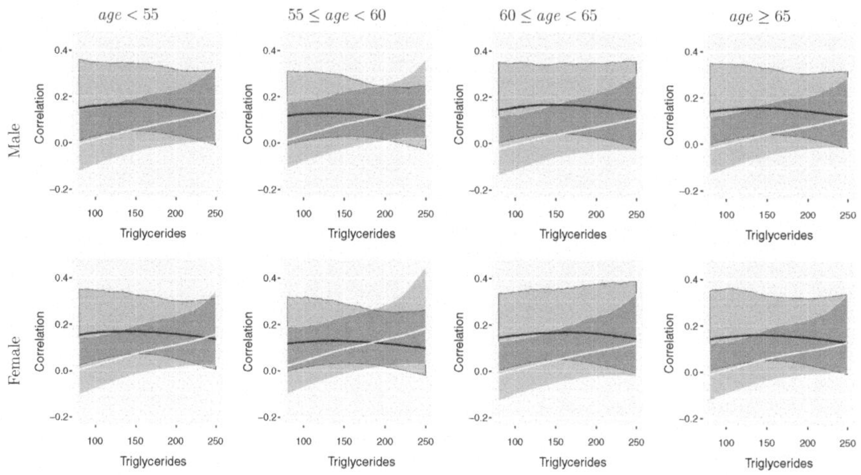

Fig. 4 Mexico City Diabetes Study Data: section plot of conditional Kendall's tau between CIMT and PG2-H (dark gray) and between CIMT and Hb1Ac (light gray) when BMI equals 32. Panels display the estimate of Kendall's tau for varying values of triglycerides, sex, and age category for normoglycemic participants: the continuous line is the estimate (dark gray for Kendall's tau between CIMT and PG2-H and light gray for Kendall's tau between CIMT and Hb1Ac) while the gray region is a 90% credible band. The y-axis shows values for Kendall's tau while x-axis shows values of triglycerides

follows approximately a standard normal distribution. For each covariate, we compute the proportion $P(\boldsymbol{x}) = \frac{1}{M}\sum_{m=1}^{M} \mathbb{1}(|S(\tau(\boldsymbol{x})^{(m)})| > z_p)$, where M denotes the number of MCMC scans, $\tau(\boldsymbol{x})^{(m)}$ denotes the m–th posterior predictive conditional Kendall's tau, $|\cdot|$ denotes the absolute value function, $\mathbb{1}(\cdot)$ denotes the indicator function, and z_p denotes the p–th quantile of the standard normal distribution, specifically the 0.975 quantile. When computing the proportion, $P(\boldsymbol{x})$, for one of the hyperglycemic markers and CIMT, the greater the proportion, the more evidence to support that hyperglycemic marker as an additional guidance for cardiovascular risk.

Figure 5 displays the proportions, $P(\boldsymbol{x})$, for Kendall's tau between CIMT and PG2-H (dark gray) and between CIMT and HbA1c (light gray) for the normoglycemic participants in the MCDS for both male and female, all categories of age, and BMI levels of 22, 27, and 32. Although the posterior predictive conditional Kendall's tau is generally small, the proportions of test statistics that are greater than $z_{0.975}$ are high for some covariate combinations. Particularly, the proportions for the test statistics of Kendall's tau between CIMT and PG2-H, for individuals with BMI equal to 32.

For individuals with BMI equal to 22, HbA1c gives more evidence of cardiovascular risk than PG2-H, uniformly for every value of the considered predictors. For individuals with BMI equal to 27, all age categories, except between 55 and 60 years, and both, male and female, PG2-H gives more evidence of cardiovascular risk for triglyceride levels lower than ~200, while for greater levels of triglycerides, HbA1c is a better cardiovascular risk marker. Interestingly, for individuals aged between 55 and 60 years, HbA1c provides more evidence of cardiovascular risk

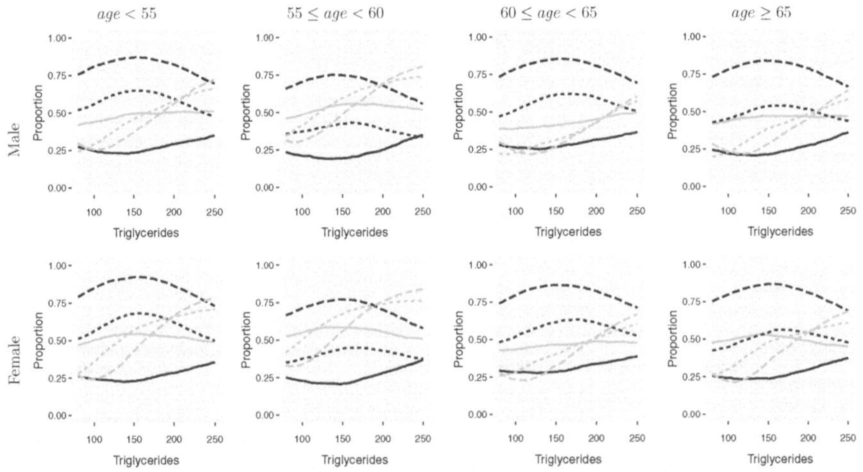

Fig. 5 Mexico City Diabetes Study Data: proportions of test statistics that are greater than $z_{0.975}$ for Kendall's tau between CIMT and PG2-H (dark gray) and between CIMT and HbA1c (light gray) for the normoglycemic participants in the MCDS for sex, all categories of age, BMI equal to 22 (solid lines), BMI equal to 27 (dotted lines), and BMI equal to 32 (dashed lines)

than PG2-H, for every triglyceride level. Finally, for individuals with BMI equal to 32, all age categories, except between 55 and 60 years, PG2-H gives more evidence of cardiovascular risk uniformly across triglyceride levels, while for individuals aged between 55 and 60, HbA1c provides more evidence of cardiovascular risk for individuals with high triglyceride levels (greater than ~200). Another interesting finding is that the use of PG2-H marker shows stronger evidence of association with CIMT as the BMI increases, uniformly across age categories and triglyceride levels, while this is observed for HbA1c only for triglyceride levels greater than ~200.

Although appealing due to the ready availability of the test, our findings indicate that the use of HbA1c as a cardiovascular risk marker is not uniformly better than a PG2-H test, across mediating covariates in the MCDS population.

6 Conclusion

Understanding T2D and its relationship with other conditions is an important and complex task. Rates of T2D are increasing worldwide and in particular, for the Mexican population. The assessment of cardiovascular risk through hyperglycemic markers is relevant. Although HbA1c is an accessible marker, our findings show that it does not provide more evidence of cardiovascular risk than PG2-H, uniformly across covariates, in normoglycemic subjects of the MCDS.

In this chapter, we have presented a general and flexible model for describing the association structure between hyperglycemic markers and cardiovascular risk, controlled by triglycerides, BMI, age, and sex. Our methodology is based on a fully BNP approach for modeling conditional copula densities. As a by-product, the conditional association structure described by the conditional Kendall's tau is obtained. Therefore, a direct understanding of this association, adjusted by known risk mediators, is feasible without the need to define thresholds in the CIMT, HbA1c, or PG2-H.

Disclosure Statement
The authors declare that there is no conflict of interest.

Data Availability Statement
Code and instructions to reproduce the results and plots for simulation Scenario I together with the dataset used in the analysis of the MCDS are provided in the github repository https://github.com/claudiawehrhahn/EvaluationOfRiskMarkersBNP.

Acknowledgments Claudia Wehrhahn was partially funded by CONACyT Project 241195, award NSF-DMS 1738053 and ATD-DMS 1441433. Ruth Fuentes-García would like to acknowledge the support of PAPIIT IN100823, UNAM. Ramsés H. Mena gratefully thanks the support of PAPIIT-UNAM project IT100524. Fabrizio Leisen acknowledges the support by the European Community's Seventh Framework Programme [FP7/2007-2013] under Grant agreement No: 630677. Maria Elena González-Villalpando and Clicerio González-Villalpando want to express their gratitude to all the participants in the community, who allowed the Mexico City Diabetes Study investigators to enter their home. This generous attitude is recognized.

References

1. F. Abegaz, I. Gijbels, and N. Veraverbeke, *Semiparametric estimation of conditional copulas*, Journal of Multivariate Analysis 110 (2012), pp. 43–73.
2. E.F. Acar, R.V. Craiu, and F. Yao, *Dependence calibration in conditional copulas: A nonparametric approach*, Biometrics 67 (2011), pp. 445–453.
3. A.F. Barrientos, A. Jara, and F.A. Quintana, *Fully nonparametric regression for bounded data using dependent Bernstein polynomials*, Journal of the American Statistical Association 112 (2017), pp. 806–825.
4. M. Burda and A. Prokhorov, *Copula based factorization in Bayesian multivariate infinite mixture models*, Journal of Multivariate Analysis 127 (2014), pp. 200–213.
5. M. De Iorio, W.O. Johnson, P. Müller, and G.L. Rosner, *Bayesian nonparametric non-proportional hazards survival modelling*, Biometrics 65 (2009), pp. 762–771.
6. M. De Iorio, P. Müller, G.L. Rosner, and S.N. MacEachern, *An ANOVA model for dependent random measures*, Journal of the American Statistical Association 99 (2004), pp. 205–215.
7. P. de Valpine, D. Turek, C.J. Paciorek, C. Anderson-Bergman, D.T. Lang, and R. Bodik, *Programming with models: writing statistical algorithms for general model structures with NIMBLE*, Journal of Computational and Graphical Statistics 26 (2017), pp. 403–413.
8. J. Escobedo-de la Peña, L. Buitrón-Granados, J. Ramírez-Martínez, R. Chavira-Mejía, H. Schargrodsky, and C. Marcet, *Diabetes in Mexico. Carmela study*, Cir. Cir. 79 (2011), pp. 424–431.
9. A.E. Gelfand, A. Kottas, and S.N. MacEachern, *Bayesian nonparametric spatial modeling with Dirichlet process mixing*, Journal of the American Statistical Association 100 (2005), pp. 1021–1035.
10. C. Genest, E. Masiello, and K. Tribouley, *Estimating copula densities through wavelets*, Insurance: Mathematics and Economics 44 (2009), pp. 170–181.
11. I. Gijbels, M. Omelka, N. Veraverbeke, *et al.*, *Multivariate and functional covariates and conditional copulas*, Electronic Journal of Statistics 6 (2012), pp. 1273–1306.
12. I. Gijbels, N. Veraverbeke, and M. Omelka, *Conditional copulas, association measures and their applications*, Computational Statistics & Data Analysis 55 (2011), pp. 1919–1932.
13. C. Gonzalez, M.P. Stern, E. Gonzalez, D. Rivera, J. Simon, S. Islas, and S. Haffner, *The mexico city diabetes study: A population-based approach to the study of genetic and environmental interactions in the pathogenesis of obesity and diabetes*, Nutrition reviews 57 (1999), pp. 71–77.
14. C. González-Villalpando, C. Dávila-Cervantes, M. Zamora-Macorra, B. Trejo-Valdivia, and M. González-Villalpando, *Incidence of type 2 diabetes in Mexico: results of the Mexico City Diabetes Study after 18 years of follow-up*, Salud Publica Mex. 56 (2014), pp. 7–11.
15. C.M. Hafner and O. Reznikova, *Efficient estimation of a semiparametric dynamic copula model*, Computational Statistics & Data Analysis 54 (2010), pp. 2609–2627.
16. J.M. Hernández-Lobato, J.R. Lloyd, and D. Hernández-Lobato, *Gaussian process conditional copulas with applications to financial time series*, in *Advances in Neural Information Processing Systems*. 2013, pp. 1736–1744.
17. Y. Huang, Y. Bi, W. Wang, M. Xu, Y. Xu, M. Li, T. Wang, Y. Chen, X. Li, and G. Ning, *Glycated hemoglobin a1c, fasting plasma glucose, and two-hour postchallenge plasma glucose levels in relation to carotid intima-media thickness in chinese with normal glucose tolerance*, The Journal of Clinical Endocrinology & Metabolism 96 (2011), pp. E1461–E1465.
18. H. Ishwaran and L.F. James, *Gibbs sampling methods for stick-breaking priors*, Journal of the American Statistical Association 96 (2001), pp. 161–173.

19. P. Lambert and F. Vandenhende, *A copula-based model for multivariate non-normal longitudinal data: analysis of a dose titration safety study on a new antidepressant*, Statistics in medicine 21 (2002), pp. 3197–3217.
20. L. Dalla Valle, F. Leisen, and L. Rossini, *Bayesian nonparametric conditional copula estimation of twin data*, Journal of the Royal Statistical Society: Series C (Applied Statistics) 67 (2017). pp. 523–548.
21. E. Levi and R.V. Craiu, *Bayesian inference for conditional copulas using Gaussian process single index models*, Computational Statistics & Data Analysis 122 (2018), pp. 115–134.
22. D. Lopez-Paz, J.M. Hernández-Lobato, and G. Zoubin, *Gaussian process vine copulas for multivariate dependence*, in *International Conference on Machine Learning*. 2013, pp. 10–18.
23. S.N. MacEachern, *Dependent Dirichlet processes*, Tech. Rep., Department of Statistics, The Ohio State University, 2000.
24. R. Neal, *Slice sampling*, The Annals of Statistics 31 (2003), pp. 705–767.
25. R.B. Nelsen, *Introduction*, in *An Introduction to Copulas*, Springer, 1999, pp. 1–4.
26. S. Ning and N. Shephard, *A nonparametric Bayesian approach to copula estimation*, Journal of Statistical Computation and Simulation 88 (2018), pp. 1081–1105.
27. A.J. Patton, *Modelling asymmetric exchange rate dependence*, International economic review 47 (2006), pp. 527–556.
28. M. Plummer, N. Best, K. Cowles, and K. Vines, *CODA: convergence diagnosis and output analysis for MCMC*, R news 6 (2006), pp. 7–11.
29. A. Rodriguez and D.B. Dunson, *Nonparametric bayesian models through probit stick-breaking processes*, Bayesian analysis (Online) 6 (2011).
30. J. Sethuraman, *A constructive definition of Dirichlet prior*, Statistica Sinica 2 (1994), pp. 639–650.
31. A. Sklar, *Fonctions de répartition à n dimensions et leurs marges. publications de l'institut de statistique de l'université de paris* (1959).
32. N. Thomas, C.Q. Jiang, S.M. McGheea, W.S. Zhang, X.Q. Lao, M. Schoolinga, P. Adab, T.H. Lam, and K.K. Cheng, *Association of vascular risk factors with increasing glycemia even in normoglycemic subjects in an older chinese population: the guangzhou biobank cohort study*, Metabolism Clinical and Experimental 55 (2006), pp. 1035–1041.
33. N. Veraverbeke, M. Omelka, and I. Gijbels, *Estimation of a conditional copula and association measures*, Scandinavian Journal of Statistics 38 (2011), pp. 766–780.
34. C. Wehrhahn, A. Rodriguez, and C. Paciorek, *Bayesian Nonparametric Mixture Models Using NIMBLE*, in *NeurIPS Workshop on Nonparametric Bayesian Models*. 2018.
35. J. Wu, X. Wang, and S.G. Walker, *Bayesian nonparametric inference for a multivariate copula function*, Methodology and Computing in Applied Probability 16 (2014), pp. 747–763.
36. J. Wu, X. Wang, and S.G. Walker, *Bayesian nonparametric estimation of a copula*, Journal of Statistical Computation and Simulation 85 (2015), pp. 103–116.
37. A. Zellner, *On assessing prior distributions and Bayesian regression analysis with g-prior distributions*, Bayesian inference and decision techniques (1986).

University Infrastructure as a Factor Influencing Entrepreneurial Intention: A Structural Equation Model

Juan Carlos Neri Guzmán, Pedro Isidoro González Ramírez, Manuel Ernesto Becerra Bizarrón, and Benedicta María Domínguez Valdez

Abstract This chapter presents the results of an investigation on entrepreneurial intention in 2014 university students in Mexico. In this chapter, the theory of planned behavior (Ajzen, The theory of planned behavior, organizational behavior and human decision processes, 50(2):179–211, ISSN 0749-5978, https://doi.org/10.1016/0749-5978(91)90020-T, 1991) was used, analyzed under a structural equation model that allows us to evaluate how the variables attitude toward entrepreneurship, subjective norms, perceived behavioral control and entrepreneurial spaces allow us to explain the intention of entrepreneur in university students. The study showed that there is a statistically significant and a strong and positive relationship between entrepreneurship spaces and the entrepreneurial intention of young people studying a career in administration and business at public universities.

Keywords Structural equation model · Theory of planned behavior · Entrepreneurial intention · Public universities

J. C. Neri Guzmán (✉)
Universidad Politécnica de San Luis Potosí, San Luis Potosí, Mexico
e-mail: carlos.neri@upslp.edu.mx

P. I. González Ramírez
Universidad Autónoma de San Luis Potosí, San Luis Potosí, Mexico
e-mail: pedro.gonzalez@uaslp.mx

M. E. Becerra Bizarrón
Universidad de Guadalajara, Guadalajara, Mexico

B. M. Domínguez Valdez
Universidad Politécnica de Tulancingo, Tulancingo, Mexico
e-mail: benedicta.dominguez@upt.edu.mx

J. A. Christen et al. (eds.), *Statistics, Society and Environment*, Springer Proceedings in Mathematics & Statistics 479, https://doi.org/10.1007/978-3-031-78401-9_11

1 Introduction

Development is one of the premises that explains and justifies human's decision to live collectively, and thanks to its stimulus, it allows them to have the necessary means to develop in a social, economic, democratic and healthy context. However, development is not fortuitous, and it is the responsibility of the State to promote it, which can be achieved through various policy instruments that range from financial and administrative to the provision of infrastructure that facilitates social or economic development. In its most general form, development can be measured through some basic indicators such as production, number of companies created, number of jobs generated, number of goods produced and exported and life expectancy of a person, among others.

With such an approach, it seems that development depends exclusively on the will of the government; however, there are many factors that affect it, such as population growth, expansion of cities, availability of resources, unmet demands of public service infrastructure, government's collection capacity, public debt, external crises and attention to extraordinary events such as natural disasters or pandemics, among many others.

Likewise, a process that has influenced development expectations has been the government's withdrawal from the market starting in the 1990s, which has caused a decrease in the State's management capacity by limiting and reducing its income, which highlights its capacity to find a way to meet the growing demands of the population and development.

Given this situation, in recent decades, the government has promoted greater participation of economic agents in the resolution and attention to the demands of society, for both the production of goods and services, giving greater participation to the business sector, which seems well intentioned. However, it has generated inequalities, especially in economic activities, where return income is differentiated and more accentuated when investment is required in remote rural areas with low demand.

This limitation of the government's management capacity becomes even more acute when it is necessary to confront other problems of national or international scope, such as poverty, insecurity, informality, pollution, phenomenon of migration, aging of the population, breakdown of balance in ecosystems or negative effects derived from climate change such as desertification or lack of drinking water.

In this sense, international organizations recognize that it is necessary to define "new roles" for citizens, governments, companies and institutions, in general, to participate in a co-responsible manner in addressing the great challenges of humanity and addressing the needs of society.

As is known, universities have served throughout history as promoters of the development of humanity and contributed to the improvement of the quality of life of the population. In this context, attention is drawn to the new roles attributed to higher education institutions that, by having professors-researchers who have mastery of knowledge, methodologies, infrastructure and advanced technologies, have been identified as ideal agents to participate in meeting the demands of society and contribute to the resolution of the great challenges of humanity and the planet.

Faced with this dilemma, higher education institutions (HEIs) that register less and less financial support are invited to participate in research activities with the social and economic sectors which not only allow them to compensate for the scarcity of resources but also participate in the resolution of its problems and the great challenges of humanity. However, this requires a revolution within them to align their educational programs, teaching staff, infrastructure and students to carry out these new functions.

2 Theoretical Framework

The policy of greater participation by society to resolve common problems and maintain habitable conditions on the planet makes sense when we recognize that governments or companies by themselves do not have the capacity to do so. The idea of collaboration is not new; important advances have been made in theoretical research such as the triple helix proposal (and more recent versions of the quadruple or quintuple helix), competitiveness models, innovation ecosystems, industrial clusters or regional innovation systems, among others, where the principle is cooperation to achieve greater competitiveness; however, they have had a predominantly industrial-business focus.

More recently, international cooperation initiatives have been carried out that seek advances which overcome major global challenges such as climate change or inequalities between countries, mainly aligned with the Sustainable Development Goals where international organizations, national governments, civil society, private sector and the academy participate. In Mexico, there is the case of the International Development Cooperation Law published in 2011, where cooperation policies are contemplated at different levels; one of them, for example, is the academy [3].

What is relevant is to recognize that to advance in addressing social needs and global problems, the participation of everyone in society is necessary. However, this participation requires local actors to adjust their roles toward a context of greater collaboration and development of services identified with attention to various economic, social and environmental problems.

In this context, higher education institutions play a relevant role in cooperation because in addition to generating and transferring knowledge, they facilitate the training of citizens in priority areas of the local economy, as well as encourage entrepreneurship in young people to start or consolidate companies.

According to UNESCO, higher education has an important responsibility in managing these challenges; this is why, as a result of a study with experts from around the world, it recognizes that a more relevant higher education is required that makes use of research to resolve complex problems and overcome inequalities in society and communities in coordination with the actors involved.

However, to achieve this, a solid and strategic link is required in accordance with educational programs and collaborative work with other HEIs to intervene in multidisciplinary proposals and long-term at the local, regional and international levels, through its basic functions, which are the generation and exchange of knowledge [55].

For this reason, it envisions a university that promotes change and as an empowering source for teachers and students who facilitate the transfer of knowledge from what is learned in higher education to what is practiced at the local, national and global level to promote change through coordination between the government and civil society. In a metaphorical sense, higher education institutions are seen as a ship and a spyglass so that, through knowledge, technology and learning, we can advance toward the future in a scientific and certain way (building bridges between the backwardness and the development).

Therefore, Acosta [1] reflects on the changes required by the society, the economy and politics, as well as institutions. In particular, he highlights the need for reform in higher education from a prospective and future approach, as an entity capable of influencing transformation, through the training of professionals and researchers, but also encouraging entrepreneurship and participation of young people and institutions in addressing the needs of society.

In this context of dynamic transformation, Tuirán [54] indicates that higher education in Mexico must improve both in its coverage (it only serves 40%) and in its relevance, to adjust the supply of professionals to the new technological demands and changes in the organization of work required by the productive sector. So educational institutions, with greater emphasis on higher level institutions, must create practical learning environments and conditions to promote the development of professional competencies, technical and socio-emotional skills such as adaptability, initiative, flexibility, creativity, teamwork, perseverance and innovation that makes it easier for professionals to find work.

This environment also leads to considering the proposals of international universities [52] as a strategy to link higher education institutions to the most advanced economies in knowledge and research that access institutional models that allow them to form global citizens who contribute significantly to society and intervene in the economy with better practices in managing production, providing services, caring for the environment and helping achieve better levels of quality of life.

Although expectations regarding institutions and the higher education system are very high, this requires overcoming two major obstacles in developing countries like Mexico: on one side, addressing traditional indicators such as quality, efficiency, coverage, equity and educational relevance and, on the other side, designing and operating a higher education management model that allows greater autonomy to consolidate an infrastructure to support entrepreneurship, a link with social-economic agents; incorporating innovation in its educational programs; and consolidating a teaching staff with a new profile that identifies with teaching, research and business management, financing and self-management mechanisms, all framed in a policy with a strategic, systemic and long-term vision.

Damián [14] recognizes the relevance of this need to update higher education institutions and highlights the characteristics of university management, the consolidation of spaces for innovation such as transfer offices, science and technology parks and the importance of the factors that influence entrepreneurship.

This philosophy of the new university or entrepreneurial university must take into account that the purpose must start from facilitating the formation of human capital so that it is capable of entering the labor market and influencing the entrepreneurship and consolidation of companies, which allows it to achieve their social mobility and at the same time influence employability and productivity, thereby contributing to local-regional development.

Some authors speak of university social responsibility [60] as that institution that recognizes that its duty responds to various interest groups (stakeholders) to whom it must be accountable and evaluate the transferred benefit. Among these actors are students, teachers, administrative staff (university community), former students, the government, companies, parents, nongovernmental organizations, higher secondary education institutions, sponsors and suppliers, as well as interest groups of economic, social and environmental benefits.

This change in higher education institutions has different antecedents; in the United States, it dates back to the approval of the Compete Law (COMPETE) in 2007, where the educational system and universities in particular have played a role increasingly decisive in promoting innovation and entrepreneurship, as well as the creation of added value for their communities, which makes them an agent of change and attention to the demands of economic agents [53].

The European Union, in turn, developed the 2020 Entrepreneurship Action Plan [18], which emphasizes a strategy to enhance Europe's competitiveness and innovation. This includes reforming educational programs to make business education a key driver for creating and consolidating companies, generating jobs and income, and ultimately fostering smart, sustainable, and inclusive growth.

In Spain, the regulation of universities attributes to them, since 2001, three major responsibilities related to the training, research or generation of new knowledge and the dissemination of knowledge, innovation, development and technological transfer to society. Thanks to this last mission, universities have an effect on the productive and socioeconomic sector that can positively influence the progress and transformation of societies. This is facilitated thanks to strategies such as the registration of patents, licenses, design of training programs and implementation of applied research agreements, among others [20].

In Mexico, the recently implemented New Mexican School model [45] also registers a focus on promoting entrepreneurship in HEIs, which is conceived as a strategy that contributes to the transformation of society by involving teachers and students through a link with families, companies, public and private institutions and social organizations. This allows giving a new meaning to learning, or what is called the territorialization of higher education, where higher education institutions, their educational processes and students connect with the community to contribute to the consolidation of its social, economic and cultural development [16].

In summary, we can identify two major trends of change in universities as a result of the current context: a strengthening of the link that allows them to participate in meeting related to needs of economic agents and, on the other hand, contribute to the training of entrepreneurs who have a positive influence on the creation and consolidation of companies as sources of jobs and income that contribute to the social and economic development of a region.

This research becomes relevant by focusing on the study of higher education institutions in Mexico and their capacity to influence the training of entrepreneurs, which is done from the perception of university students. According to figures from the Association of Entrepreneurs of Mexico [5], 52% of new ventures in the country are created by people with higher education (bachelor's degree) and 81% by people with a postgraduate degree (master's or doctorate). Likewise, 43% of those who created a business did so to seek personal and professional growth and 30% to identify a profitable business opportunity. Also, it identifies that 61% of the ventures are carried out by men, and 41.4% of the ventures are created by people with an average age of 26 and 35 years. These data show the relevance of higher education to train professionals who influence entrepreneurship; so this study becomes important to consolidate this culture.

Different studies show that education with a focus on entrepreneurship affects the attitude of students and is related to business success [6, 11, 15, 26, 35, 56]. Therefore, it is important to verify if in universities educational programs, infrastructure and support policies exist that facilitate the development of entrepreneurial attitudes in students, with greater emphasis on higher education institutions, which places professionals directly into the labor market.

For the European Commission [26], entrepreneurship education considers all those educational activities that have the purpose of preparing responsible and entrepreneurial people who have the skills, knowledge and attitudes necessary to achieve the goals they set to live a full life.

The study of entrepreneurship has been carried out from different approaches, both due to the characteristics of higher education institutions [41] and the perception of students. In both cases, there are advances, especially theoretical, and some others include statistical models that allow evaluating the effective relational relationships between the factors that affect entrepreneurship. Research on entrepreneurship from university students has mainly had the following focuses:

Authors such as Rubio and Lisbona [43] found that scientific studies on entrepreneurship have been carried out with more intensity in the last decade and that the most recurrent theoretical approaches are the theory of planned behavior and the entrepreneurial event model. Likewise, they found that 43% of the data analysis techniques applied in entrepreneurship studies are related to those that make use of statistical validity, of which 52% use structural equation models. Cortez and Hauck [13] found similar results, identifying that four out of ten studies on entrepreneurial intention used the theory of planned behavior as a reference, all using reasonable reliability and validity indices to demonstrate their theoretical consistency.

The theory of the entrepreneurial event [48] is based on the fact that the intention to start a venture is the result of the calculation of the perceived convenience and the possibility of becoming an entrepreneur. So it is about choosing an option of business that is most viable (convenience and viability) and with the greatest possibility of success [51].

The theory of planned behavior [2] focuses on explaining a person's intention to carry out some activity, action or objective, through its determining factors: the attitude toward the behavior, subjective norms and the perceived behavioral control. However, different authors have enriched this theory by adding dimensions of analysis that allow it to be enriched when the topic of entrepreneurship is addressed, such as self-efficacy, motivation, need for achievement, propensity to take risks, need for independence and ability to innovate, mainly [30, 34], which for the most part respond to personal and interpersonal aspects. This approach is usually validated through the use of multivariate statistics and confirmatory statistical analyses, or structural equation models.

This research takes up the theory of planned behavior to measure the perception of entrepreneurship that students have in public universities, considering an additional aspect: entrepreneurship spaces, a dimension that has been little studied and that we consider has great relevance for the consolidation of entrepreneurship in young university students, taking into account that it represents the physical, administrative, financial, teaching environment, educational programs, subjects and activities—conditions that favor creating an environment fostering entrepreneurial culture in which the student is exposed throughout the process of vocational training. Some authors [25, 36] mention that interpersonal, contextual and infrastructure characteristics can have a greater impact on entrepreneurial intention than personal factors themselves, such as the environment.

The theoretical approach used is validated with the use of the structural equation model or structural equation models (SEM), as a highly reliable tool that is increasingly used in social sciences as it allows verifying the consistency between the relationships. Existing between the theoretical dimensions of the study and the observed variables that are used to measure them. Through the "Scopus" platform, Soriano and Mejía [49] generated a search for research where structural equation models are used, finding that 11.4% of them are applied to studies on business and administration and that this tool has charged interest and application more significantly from the year 2000.

Therefore, a good start in researching is to know if we have institutions that promote entrepreneurship. Or if there are incentives that favor creativity and entrepreneurship in higher education institutions?

3 Method

This research aims to carry out a study to verify how the infrastructure is promoting entrepreneurship in public higher education institutions in Mexico and how it affects the entrepreneurial intention of students. In order to do this, we start from the theory of planned behavior and propose a structural equation model to evaluate the consistency of the theoretical approach.

The specific objectives are using the theory of planned behavior to measure the entrepreneurial characteristics of university students; validating the fit of the theoretical entrepreneurship model for the local case through a structural equation model; and verifying that the infrastructure in public universities in Mexico has an effect on the entrepreneurial intention of students.

This way, the research questions that need to be answered are the following: Is there an influence between the determinants and entrepreneurial intention in students from public universities in Mexico? Is it possible to verify the consistency of the theoretical model of entrepreneurial intention through a structural equation model? And does the entrepreneurship infrastructure significantly and positively affect the entrepreneurial intention of university students?

The study is based on a survey of 2014 university students in careers related to administration and business because it is considered that they are the educational programs where the development of skills to create entrepreneurship is promoted with greater emphasis or if the students have the necessary tools to contribute to the consolidation of companies.

The methodology that underlies this research is the theory of planned behavior [2], which indicates that it is possible to predict people's intention to carry out some activity, with a high level of reliability, if people's attitudes, subjective norms and perceived control behavior are taken into account. This study additionally includes a dimension of analysis that is the entrepreneurship infrastructure, considering that, to facilitate the development of entrepreneurship, it is necessary that higher education institutions have an infrastructure (physical, administrative, human and financial) and appropriate strategies that facilitate the transfer of this competence to their students. Therefore, the hypothesis that we seek to verify is to show that the entrepreneurship infrastructure has a positive and significant impact on the entrepreneurial intention of public higher education students in Mexico.

To support the research, structural equation models or structural equations models (SEM) are used as a flexible technique that allows modeling complex relationships between variables that are observed, called observables, and variables that are not manifested, identified as latent [27]. In this research, the analysis variables are defined as follows:

Entrepreneurial intention. Take into account the motivational factors that influence the student's entrepreneurial behavior, as a dependent variable.

Attitude toward entrepreneurship. It reflects the degree to which a student values entrepreneurial behavior, as an independent variable.

Subjective norms. It refers to the social pressure perceived by the student to show entrepreneurial behavior, as an independent variable.

Perceived behavioral control. It shows the student's confidence in demonstrating entrepreneurial behavior and maintaining control of the factors that determine it, as independent variables.

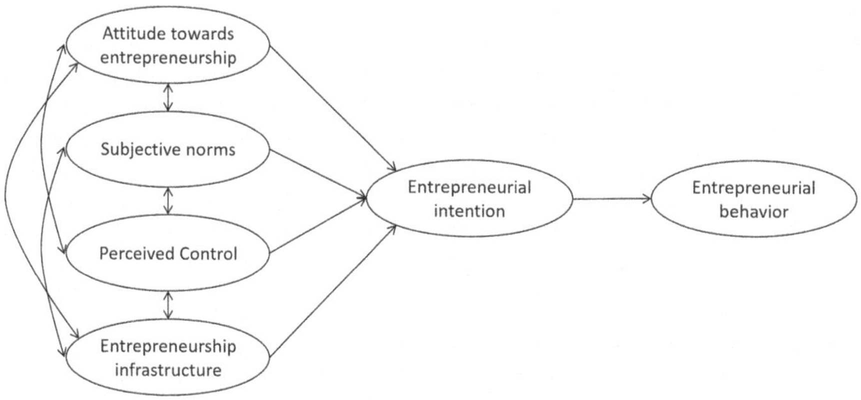

Fig. 1 Theoretical model of entrepreneurial intention in university students

Entrepreneurship infrastructure. Consider the spaces available at a higher education institution to encourage the generation of ideas, development of creativity and entrepreneurial spirit of their students, as independent variables.

The study of entrepreneurial intention is analyzed through an SEM model which has among its advantages quantifying the effect and the relationship that exists between variables called latent, or large theoretical constructs, which cannot be measured directly, which in this research are attitude toward entrepreneurship, subjective norms, perceived behavioral control, entrepreneurial infrastructure and entrepreneurial intentions (the first four represent the independent variables and the last one the dependent variable), and the observed variables, which can be measured directly, which in our case are represented by a total of 25 variables and are used to estimate the first ones.

The SEM represents a highly useful tool to verify that the proposed theoretical approach is consistent with the observed one, which is one of the objectives of this research, and is carried out through a multivariate statistical analysis that contrasts the variance and covariance matrices, which shows the relationship between variables through correlation and association indices. An additional advantage of using the SEM is that it allows graphically representing and measuring the relationship between the observed and latent variables, as well as the relationship between latent variables, including showing the error variances, thus providing greater capacity for analysis and reliability in the results, compared to other mathematical-statistical methodologies available [4, 37, 42, 46].

Graphically, in this research, the theoretical model represents the observed variables with rectangles, the latent variables with ellipses, the correlation with bidirectional arrows, the association between them with unidirectional arrows and the estimation errors in smaller ellipses (Fig. 2).

Although the use of SEM has come to support the causal-relational theoretical approaches in the social sciences, one of the critical aspects in its use is to support the assumption of normality of the data, particularly when the sample size is large (as in this research), because it increases the differences between the variance-covariance matrices. The recommended sample size is greater than 250 and the number of cases is between 10 and 20 times the number of variables; in this case, the ratio is 1:77 [7, 8, 33, 37, 44, 50]. As this condition is very common in research, affecting hypothesis testing and absolute goodness-of-fit tests, it is recommended to consider tests with some incremental goodness-of-fit statistics and parsimony indicators at the same time [23, 24, 28, 31, 32, 40, 47, 57, 59].

In this sense, this research evaluates the consistency of the structural equation model of entrepreneurial intention considering some of the three types of indicators that allow estimating the statistics of goodness-of-fit statistics and quality of the relationship of its variables (latent and observed).

(a) The absolute fit indices are based on the assessment of the residuals and the variance is not explained by the model per degree of freedom. The most used statistics are the chi-square and the root mean squared error of approximation (RMSEA). Root means error of approximation allows us to test the hypothesis in which the model errors are null; however, it is difficult to fulfill it because they are very sensitive to the size of the sample, even when the model has a good fit. The relationship between chi-square and degrees of freedom is expected to take values between 1 and 5, while RMSEA between 0.5 and 0.8. The root mean square residual (RMR). The RMR Index evaluates the differences between the variance and covariance matrices of the sample and the model; when the indicator is less than 0.05, it is considered a good adjustment [19].
(b) Incremental adjustment indices. The incremental adjustment indices allow us to compare the proposed model with the independence model, where it is assumed that there is no association between the variables, evaluating the improvement of the former with respect to the second one. The most used indicators are the comparative fit index (CFI), the goodness-of-fit index (GFI), the incremental fit index (IFI) the Tucker-Lewis index (TLI) and the normalized fit index (NFI), which take values between 0.9 and 1 [10, 23, 29, 38].
(c) Parsimony adjustment indices. These indices, unlike absolute fit indices, take into account the degrees of freedom as a weighting factor and are more useful for complex models. The most used are the parsimony-goodness-of-fit index), the parsimony normed fit index (PNFI) and the parsimony comparative fit index (PCFI), based on the GFI, NFI and CFI. These indices assess a good fit for results greater than 0.5 [17, 39].

The study of entrepreneurship takes into account five latent variables: attitude toward entrepreneurship, subjective norms, perceived behavioral control, entrepreneurial intentions and entrepreneurship infrastructure, which are studied through 5, 4, 6, 5 and 5 observed variables, respectively. These are last measured on a Likert scale with five ranges of agreement: Not at all agree, Somewhat agree, Quite a lot agree, Strongly agree and Total agreement, which were interpreted quantitatively based on the degree of agreement.

To validate the relevance and theoretical assumptions of the model, a confirmatory factor analysis (CFA) of the latent variables with respect to the observed variables was carried out, which showed good reliability measured through a McDonald's Omega statistical test, which is recommended when we have nominal variables valued with a Likert scale in five levels and that have greater reliability due to working with factor loadings of the standardized variables [21, 22, 58], with the following coefficients 0.878, 0.840, 0.862, 0.910 and 0.942 for each component: attitude toward entrepreneurship, subjective norms, perceived behavioral control, entrepreneurial intentions and entrepreneurship infrastructure, respectively. To estimate and evaluate the structural model, the Analysis of Moment Structures (AMOS) statistical package of SPSS v. 25 was used.

4 Results

The structural equation model showed the correct adjustment in explaining entrepreneurial intention with respect to the variables considered by the theory of planned behavior: entrepreneurial attitude, subjective norms and behavioral control, as well as that considered on the entrepreneurial infrastructure in public universities.

Although the absolute chi-square indicator and its adjustment with respect to the degrees of freedom were not met, the rest of the indicators show an acceptable goodness of fit, which allows us to verify that the theoretical model is theoretically relevant and statistically relevant. The evidence shows that the relationship between the observed and latent variables, as well as the relationship between latent variables, are consistent, with absolute fit indices of X2 (289) = 3028.186 with $p < 0.05$, RMSEA = 0.068, RMR = 0.051; relative fit indices of CFI = 0.92, IFI = 0.92, TLI = 0.91, GFI = 0.883; NFI = 0.913 and parsimony fit indices PGFI = 0.727, PNFI = 0.812 and PCFI = 0.818.

Figure 2 shows in large ellipses the five latent variables studied (the relationship between them is known as the structural model), which are measured through the 25 observable variables, represented in rectangles; the small ellipses show the errors associated with the observed variables with respect to their corresponding latent variables, the bidirectional arrows that relate the latent variables show the correlations between them; in our case, they all represent significant values ($p < 0.001$).

To interpret the correlation between variables, Cohen's correlation scale was considered [12], where a value less than 0.1 is considered null, between 0.1 and 0.3 weak, between 0.3 and 0.5 moderate and greater than 0.5 considered strong. Therefore, the model showed that entrepreneurial intention shows a strong and significant relationship with entrepreneurial attitude, behavioral control and entrepreneurial spaces (0.811, 0.757, 0.576, respectively and $p < 0.001$), however, its relationship with subjective norms shows a significant but moderate relationship (0.35 and $p < 0.001$) (Fig. 2).

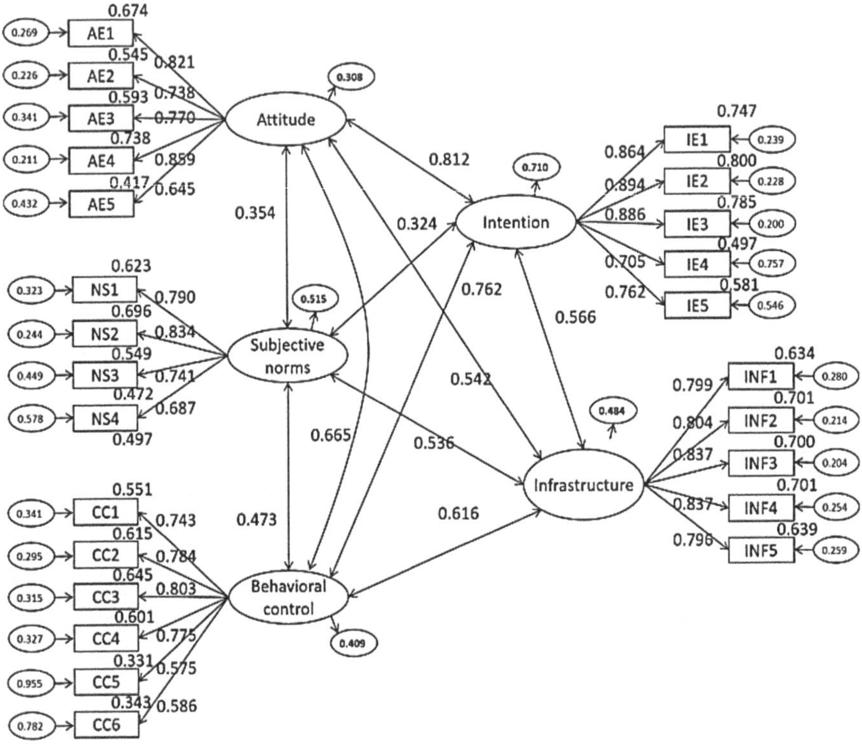

Fig. 2 Structural model of entrepreneurial intention in university students

Meanwhile, the squared multiple correlations shown in the structural model of entrepreneurial intention in Fig. 2 show, for each of the observed variables, the variance that is explained by the corresponding latent variable (value in the upper right part of rectangles), which are generally high, where 77% of the values are greater than 0.5, which is known as the reliability of the observed variables. For example, in the case of the observable variable AE1 (I am interested in becoming an entrepreneur), 67.4% of the variance is explained by the latent variable attitude toward entrepreneurship, which means that the remaining 32.6% of its variance is explained by the measurement error factor.

On the other hand, the smaller ellipses represent the measurement errors of the observable variables versus the latent variables (non-shared variances), they represent both the errors associated with the measurement of a variable and the set of variables that have not been contemplated in the model, which represents the prediction error [9, 23].

Regarding the latent variables, the error is interpreted as the variance explained by the rest of the latent variables, for example, 71% of the variance of F5 (entrepreneurial intention) is explained by the latent variables F1 (attitude toward entrepreneurship), F2 (subjective norms), F3 (behavioral control) and F4 (entrepreneurship infrastructure).

5 Conclusions

This study allowed us to validate the consistency of the theory of planned behavior to explain the entrepreneurial intention of university students, verifying the relationship of both the original dimensions: attitude toward entrepreneurship, subjective norms, perceived behavioral control and, the one theoretically proposed in this research, entrepreneurship infrastructure, which was verified with the use of the structural equation model, which allowed through its goodness-of-fit indicators and significance levels for the relationship between variables.

While determining which factors have the most influence on entrepreneurial intention, a significant and positive relationship between three factors was shown: attitude, behavioral control and entrepreneurship infrastructure. In this research, subjective norms did not show a significant relationship.

The results of this research indicate that the infrastructure or spaces to promote entrepreneurship is a necessary aspect to consolidate the entrepreneurial attitude of university students, which is why it is important to consider it within the policy of higher education institutions.

Suggestions

To obtain a broader vision of the role of infrastructure as a factor to encourage entrepreneurship, it is recognized that it would be interesting to extend it to university students from other areas of knowledge. Likewise, recognizing that entrepreneurship is a training that is consolidated throughout professional life at a university. We consider that it would be of great value to carry out studies in different semesters that allow us to show the consolidation of entrepreneurial skills and potential. Likewise, it would be relevant to carry out ex post studies to know the degree of entrepreneurship in graduates or to validate their entrepreneurial culture.

References

1. Acosta, A. El futuro de la educación superior en México. Revista Iberoamericana de Educación Superior. Núm. 13 Vol. (2014). http://ries.universia.net
2. Ajzen, I. The theory of planned behavior, Organizational Behavior and Human Decision Processes, Volume 50, Issue 2, 1991, Pages 179–211, ISSN 0749-5978, (1991). https://doi.org/10.1016/0749-5978(91)90020-T.
3. Amexid. Agencia Mexicana de Cooperación Internacional para el Desarrollo. (2023). https://www.gob.mx/amexcid
4. Arbuckle, J. L. Amos Users' Guide. Version 27. Amos Development Corporation. (2020).
5. ASEM. La Radiografía del Emprendimiento en México. (2022). https://asem.mx/

6. Bell, J.R., Dearman, D., y Wilbanks, J.E. Evaluating the effects of a problem-based learning business planning course on student entrepreneurial intentions. Journal of Entrepreneurship Education, 18(1), 169–182. (2015).
7. Bentler PM, Yuan K. Structural equation modeling with small samples: Test statistics. Multivariate Behavioral Research. 34:181–197. (1999).
8. Boomsma A., Hoogland J. J. The Robustness of LISREL Modeling Revisited. (2001). https://www.scirp.org/reference/referencespapers?referenceid=2345905
9. Byrne, B. M. Structural equation modeling with AMOS: basic concepts, applications, and programming. Routledge. Taylor & Francis Group. (2010).
10. Cho, G., Hwang, H., Sarstedt, M., & Ringle, Ch. M. Cutoff criteria for overall model fit indexes in generalizedstructured component analysis. Journal of Marketing Analytics. (2020). https://doi.org/10.1057/s41270-020-00089-1
11. Civera, A., Meoli, M., & Vismara, S. Engagement of academics in university technology transfer: Opportunity and necessity academic entrepreneurship European Economic Review, 123. (2020). https://www.sciencedirect.com/science/article/abs/pii/S0014292120300088
12. Cohen, J. Statistical Power Analysis for the Behavioral Sciences (2nd Edition). Lawrence Erlbaum Associates. (1988).
13. Cortez, P. A. & Hauck, N. Revisão de literatura de instrumentos de avaliação de intenção empreendedora. Cuadernos Latinoamericanos de Administración, 16(30), 1–17. (2020). https://doi.org/10.18270/cuaderlam.v16i30.3023
14. Damián S., J. El emprendimiento en el contexto de la universidad no emprendedora: la voz de los estudiantes. Innovación Educativa, ISSN: 1665–2673, vol. 20, número 84. (2020).
15. Díaz G., C., Sáez M., F., y Jiménez M., J. Evaluación del impacto del programa educativo "Emprendedores" en la intención emprendedora de los participantes. RUSC. Universities and Knowledge Society Journal, 12(3), 17–31. (2015). https://doi.org/10.7238/rusc.v12i3.2146
16. DOF. Ley General de Educación Superior. Cámara de Diputados del H. Congreso de la Unión. (Art. 8, Frac. XXII). (20/04/2021).
17. Doral, F. F., Rodríguez, A. I., Meseguer, A. A. Modelos de ecuaciones estructurales en investigaciones de ciencias sociales: Experiencia de uso en Facebook. Revista de Ciencias Sociales (Ve), vol. XXIV, núm. 1, pp. 22–40. (2018). https://www.redalyc.org/journal/280/28059578003/html/
18. EC. Entrepreneurship 2020 Action Plan. Reigniting the entrepreneurial spirit in Europe. (2013). https://eur-lex.europa.eu/legal-content/EN/TXT/PDF/?uri=CELEX:52012DC0795
19. Escobedo M, Hernández J, Estebané V, Martínez G. Modelos de Ecuaciones Estructurales: Características, Fases, Construcción, Aplicación y Resultados. Ciencia y Trabajo. 18 [55]: 16–22. (2016).
20. Escribá E., A., Iborra J., M., Safón, V. Modelos de dirección estratégica en universidades españolas de alto desempeño. Bilbao: Fundación BBVA. (2019).
21. Hancock, G. R., & An, J. A closed-form alternative for estimating omega reliability under unidimensionality. Measurement: Interdisciplinary Research and Perspective, 18, 1–14. (2020).
22. Hayes, A. F., & Coutts, J. J. Use omega rather than Cronbach's alpha for reliability. But…, Communication Methods and Measures, 14, 1–24. (2020).
23. Herrero, J. El Análisis Factorial Confirmatorio en el estudio de la Estructura y Estabilidad de los Instrumentos de Evaluación: Un ejemplo con el Cuestionario de Autoestima CA-14. Intervención Psicosocial. Vol. 19, n. 3, 2010—pp. 289–300. (2010). https://doi.org/10.5093/in2010v19n3a9
24. Hooper, D., Coughlan, J. and Mullen, M. R. "Structural Equation Modelling: Guidelines for Determining Model Fit." The Electronic Journal of Business Research Methods Volume 6 Issue 1, pp. 53–60 (2008). https://academic-publishing.org/index.php/ejbrm/article/view/1224
25. Huyghe, A., & Knockaert, M. The influence of organizational culture and climate on entrepreneurial intentions among research scientists. The Journal of Technology Transfer, 40(1), 138–160. (2015). https://doi.org/10.1007/s10961-014-9333-3
26. IFC. Entrepreneurship Education: A road to success A compilation of evidence on the impact of entrepreneurship education strategies and measures. European Commission. (2015).

27. Igolkina, A, & Meshcheryakov, G. Semopy: A Python Package for Structural Equation Modeling. Structural Equation Modeling: A Multidisciplinary Journal, 27(6), 952–963, (2020). https://doi.org/10.1080/10705511.2019.1704289
28. Jackson, D. L. Revisiting sample size and number of parameter estimates: Some support for the N:q hypothesis. Structural Equation Modeling, 10, 128–141. (2003).
29. Jordan M, F. M. Valor de corte de los índices de ajuste en el análisis factorial confirmatorio. Revista de Investigación en Psicología Social. vol. 7, núm. 1. (2021). http://portal.amelica.org/ameli/jatsRepo/123/1232225009/index.html
30. Karimi, S., Biemans, H., Lans, T., Arasti, Z., Chizari, M., Mulder, M. Application of Structural Equation Modelling to Assess the Impact of Entrepreneurial Characteristics on Students' Entrepreneurial Intentions. H. Fulford (Eds.). Proceedings of ECIE 2011. The 6th European Conference on Entrepreneurship and Innovation, Aberdeen, Scotland, UK, pp. 954–967. (2011).
31. Kline, R.B. Principles and Practice of Structural Equation Modeling (2nd Edition ed.). New York: The Guilford Press. (2005).
32. Lai, K. Fit Difference Between Nonnested Models Given Categorical Data: Measures and Estimation. Structural Equation Modeling: A Multidisciplinary Journal. (2020). https://doi.org/10.1080/10705511.2020.1763802
33. Lewis, T. F. Evidence Regarding the Internal Structure: Confirmatory Factor Analysis. Measurement and Evaluation in Counseling and Development. (2017). https://doi.org/10.1080/07481756.2017.1336929
34. Liñán, F.; Chen, Y. W. Testing the entrepreneurial intention model on a two-country simple. (2006). https://www.researchgate.net/publication/28117836
35. Liñán, F. y Fayolle, A. A systematic literature review on entrepreneurial intentions: Citation, thematic analyses, and research agenda. International Entrepreneurship and Management Journal, 11(4), 907–933. (2015). https://doi.org/10.1007/s11365-015-0356-5
36. Llouga, S.N., Mouloungni, A.N., & Sahut, J.M. Entrepreneurial intention and career choices: The role of volition. Small Business Economics, 42(4), 717–728. (2014). https://doi.org/10.1007/s11187-013-9524-6
37. Manzano P., A. P. Introducción a los modelos de ecuaciones estructurales. Investigación en Educación Médica, UNAM, vol. 7, núm. 25, pp. 67–72. (2018). https://doi.org/10.1016/j.riem.2017.11.002
38. McNeish, D., Stapleton, L. M., & Silverman, R. D. On the unnecessary ubiquity of hierarchical linear modeling. Psychological Methods, 22(1), 114–140. (2017). https://doi.org/10.1037/met0000078
39. Mulaik, S. A., James, L. R., Van Alstine, J., Bennet, N., Lind, S., and Stilwell, C. D. "Evaluation of Foodness-of-Fit Indices for Structural Equation Models", Psychological Bulletin, 105, 430–445. (1989).
40. Muthén, L.K. and Muthén, B. (2001) Mplus: Statistical Analysis with Latent Variables, User's Guide. Muthén & Muthén, Los Angeles.
41. Neri Guzmán, J. C. "Universidades públicas en México: análisis del emprendimiento" en Problemas del Desarrollo. 55 (216), p. 141–168. (2024). https://doi.org/10.22201/iiec.20078951e.2024.216.70026
42. Rojas T., L. Robustez de los índices de ajuste del análisis factorial confirmatorio a los valores extremos. Revista de Matemática: Teoría Y Aplicaciones 27(2). 383–404. (2020). https://doi.org/10.15517/rmta.v27i2.33677
43. Rubio H., F. J. y Lisbona B., A. M. Intención emprendedora en estudiantes universitarios. Revisión sistemática de alcance de la producción científica. Universitas Psychologica, vol. 21, pp. 1–22. (2022). https://doi.org/10.11144/Javeriana.upsy21.ieeu
44. Ruiz, M. A., Pardo A. San Martín, R. Modelos de ecuaciones estructurales. Papeles del Psicólogo, 2010. Vol. 31(1), pp. 34–45. (2010). http://www.cop.es/papeles
45. SEP. La Nueva Escuela Mexicana (NEM): Orientaciones para padres y comunidad en general. Secretaría de Educación Pública. Subsecretaría de Educación Media Superior. (2023).

46. Schermelleh E., K., Moosbrugger, H., Müller H. Evaluating the Fit of Structural Equation Models: Tests of Significance and Descriptive Goodness-of-Fit Measures. Methods of Psychological Research Online 2003, Vol.8, No.2, pp. 23–74. (2003).
47. Schumacker RE, Lomax RG. A begginer's guide to structural equation modeling. 2nd ed. Mahwah, NJ: Lawrence Erlbaum Associates. (2004).
48. Shapero, A. and Sokol, L. The Social Dimensions of Entrepreneurship. In: Encyclopedia of Entrepreneurship, Prentice-Hall, Englewood Cliffs, 72–90. (1982). https://papers.ssrn.com/sol3/papers.cfm?abstract_id=1497759
49. Soriano, J. L. y Mejía T., J. Structural Equations Modeling in the Management Sciences. Revista de métodos cuantitativos para la economía y la empresa (33). Páginas 242–263. (2022). www.upo.es/revistas/index.php/RevMetCuant/article/view/5414
50. SS. Structural Equation Modeling. Complete Dissertation by Statistics Solutions. (2022). https://www.statisticssolutions.com/free-resources/directory-of-statistical-analyses/structural-equation-modeling/
51. Tarrats, E., Torras, M. M., & Hernández, X. F. Del modelo del evento emprendedor al modelo sistémico de emprendimiento. Empresa, 4(2), 124–135. (2015). https://doi.org/10.17993/3cemp.2015.040122.124-135
52. Téllez M., F. A. y Vera A., S. M. Internacionalización y educación superior: una apuesta por la formación de ciudadanos globales. En Desafíos actuales de la educación superior: análisis y perspectivas frente a un mundo cambiante. Briceño Martínez, J.J., Castellanos Saavedra, M.P. y Valencia Molano, J. L. (2020).
53. The America COMPETES Act of 2022. https://reason.com/wp-content/uploads/2022/01/America-COMPETES-Act-of-2022-HR-4521.pdf
54. Tuirán, R. De nada sirve ampliar los niveles de cobertura si no lo hacemos con calidad. Revista Forum. Foro Consultivo Científico y Tecnológico, AC. No 38. (2018).
55. UNESCO. Caminos hacia 2050 y más allá. Resultados de una consulta pública sobre los futuros de la educación superior. Organización de las Naciones Unidas para la Educación, la Ciencia y la Cultura, y el Instituto Internacional de la UNESCO para la Educación Superior en América Latina y el Caribe (IESALC). (2021). www.unesco.org/open-access/terms-use-ccbysa-sp.
56. Valencia, A., Montoya, I., y Montoya, L. Factores explicativos de las intenciones emprendedoras en estudiantes universitarios. Espacios, 36(05), 1–15. (2016).
57. Vázquez M., J. Modelos de ecuaciones estructurales en Psicología. Tesis de Maestría. Universidad Politécnica de Valencia. (2013). http://hdl.handle.net/10251/44523
58. Ventura L., J. L; Caycho R., T. El coeficiente Omega: un método alternativo para la estimación de la confiabilidad. Revista Latinoamericana de Ciencias Sociales, vol. 15, núm. 1, enero-junio, 2017, pp. 625–627. (2017).
59. Xia, Y., & Yang, Y. RMSEA, CFI, and TLI in structural equation modeling with ordered categorical data: Thee story they tell depends on the estimation methods. Behavior Research Method, 51, 409–428. (2019). https://doi.org/10.3758/s13428-018-1055-2
60. Ying J. H., Yenchun J., W., Lan Y., H and Chia F.C, Management and Administration of Higher Education Institutions in Times of Change, 47–56. (2019).

An Approach of Latent Class Mixed Models for Analyzing New Genotypes of *Saccharum spp*

Carlos David Carretillo-Moctezuma, María Guzmán-Martínez, Eduardo Pérez-Castro, Flaviano Godínez-Jaimes, and José C. García-Preciado

Abstract The demand for new sugarcane varieties to meet market demand is constant in the sugar industry. According to the National Committee for the Sustainable Development of Sugarcane, Mexico has problems with efficiency and profitability, as it continues to fail to renew and diversify the commercial varieties it uses to grow sugarcane, which translates into more than 70% of the area dedicated to sugarcane cultivation. This generates significant economic losses. In addition, there is the demand for food and the optimization of natural resources for food production. The goal of this work is to analyze the performance of 34 new *Saccharum spp* hybrids and six commercial varieties, in the adaptability test phase in the plant cycle, using latent class mixed models. Six industrial quality variables were used for the study: sucrose, brix degrees, purity, reducing agents, fiber, and humidity. According to the results, only the fiber percentage separates the materials into two groups, the group with materials of higher fiber percentage, which includes the commercial varieties of early and early-intermediate maturity; the group with materials of higher fiber percentage, which consists of the commercial varieties of early and early-intermediate maturity; and the group with materials containing a lower rate of fiber, which includes the commercial variety of intermediate-late maturity.

Keywords Latent groups · Longitudinal data · Genotypes · Industrial quality variables · Plant cycle

---

C. D. Carretillo-Moctezuma (✉) · M. Guzmán-Martínez · F. Godínez-Jaimes
Facultad de Matemáticas, Universidad Autónoma de Guerrero, Chilpancingo de los Bravo, Guerrero, Mexico
e-mail: fgodinezj@uagro.mx

E. Pérez-Castro
Unidad de Investigación de Salud en el Trabajo, Centro Médico Nacional Siglo XXI, Instituto Mexicano del Seguro Social, Ciudad de México, Mexico

J. C. García-Preciado
Campo Experimental Tecomán, Instituto Nacional de Investigaciones Forestales, Agrícolas y Pecuarias, Mexico
e-mail: garcia.concepcion@inifap.gob.mx

J. A. Christen et al. (eds.), *Statistics, Society and Environment*, Springer Proceedings in Mathematics & Statistics 479, https://doi.org/10.1007/978-3-031-78401-9_12

1 Introduction

The sugar industry, a vital component of the agricultural economy in several regions of Mexico [1], faces a challenge of considerable magnitude related to the efficient identification and classification of sugarcane hybrids and varieties. This challenge consists of knowing the optimum time to harvest the sugarcane, as this ensures that the maximum sucrose concentration is obtained [2]. To determine the optimal harvest time, it is necessary to know the type of maturity of the genotype. Sugarcane maturity is defined as the final stage of the physiological process that results in the maximum accumulation of sucrose in the plant [3]. Knowledge of the type of maturity of the materials allows the optimization of economic and natural resources, as well as the correct management of the crop [4, 5]. Among the types of maturity of commercial varieties are: early, early-intermediate, intermediate, intermediate-late, and late [5, 6].

The identification of new commercial varieties for the sugar industry is influenced by environmental, edaphic, and agronomic factors [4], the type of genetic material, the crop cycles (plant, ratoon, and stubble), the stages of material development (testing, adaptation, agro-industrial, and semi-commercial), and the agronomic and industrial quality variables of the genotypes [7]. Before moving on to the adaptation phase, new hybrids are subjected to various testing phases for an average period of 12 years [8]. Once this phase is completed, the materials move on to the last three advanced stages of selection: adaptation, agro-industrial, and semi-commercial.

In the adaptability phase, the goal is to evaluate the new hybrids under different soil and climatic conditions [9]. The variables of interest at this stage are agronomic and industrial quality. Maturity studies at this stage are often limited to graphical analyses, which generally do not guarantee the reliability of the conclusions obtained.

In the agro-industrial phase, the maturity curve of the new hybrids is studied, and the type of maturity is determined. Generally, the variables used for this study are sucrose percentage, brix degrees, and polarization percentage [10, 11]. The study of the maturity curve is carried out using polynomial regressions, where the sucrose percentage is the dependent variable and the evaluation date is the independent variable [12], or simple linear regressions [13]. Other studies examine the maturity curve and the degree of maturity using graphical comparisons of second-degree polynomials [14, 15]. Some research has identified the materials with the highest field yield and sucrose percentage through analysis of variance [16, 17]. Multivariate clustering analyses have also been employed [18]. Recently, linear mixed models (LMM) have been utilized [19]. This method, unlike the previous ones, is the most suitable for analyzing repeated measurements over time (longitudinal data) of genotypes. However, it does not generate a natural grouping of materials, which makes it difficult to analyze various growth curves to identify the maturity of the materials.

However, combinations of various statistical methods have also been used to study the maturity curve of the materials [19]. It is necessary to continue exploring new statistical tools to assist in the optimal identification of the best-performing materials in the field.

A statistical tool to generate a natural grouping of the data is the latent class mixture model (LCMM), which combines the characteristics of LMMs and latent class models. The latter is mainly aimed at identifying unobserved classes or groups that differ qualitatively and represent patterns of association between observations. With this family of models, it is possible to study latent trajectory profiles in heterogeneous populations, considering the correlation between repeated measurements for the same subject and distinguishing between latent groups. LCMMs have primarily been used in the social and health sciences to classify individuals or objects into different groups/classes based on responses to a set of observed variables. In agronomy, however, the use of LCMMs is limited; LCMMs have been employed to describe patterns of improved agrotechnology use in Uganda [20] and farmers' preferences regarding filter strips programs [21]. LCMMs are valuable because they handle heterogeneity in observations and assess patterns of change over time [22].

Given the economic importance of sugarcane cultivation at the national and international levels, it is imperative to continue studying *Saccharum spp.* genotypes under current environmental conditions. This effort aims to develop new commercial varieties capable of meeting market demand and adapting to environmental changes.

The goal of this work is to analyze the behavior of 34 new sugarcane hybrids that are in the adaptability test phase in the plant cycle and to compare their performance with that of six commercial varieties. Six industrial quality variables and LCMMs will be used for the analysis.

2 Materials

The study was conducted at the Melchor Ocampo sugar mill, located in Colonia Predio Zacapala, Autlán de Navarro, Jalisco, at 19°47′18.0″ north latitude and 104°14′24.3″ west longitude. The materials were established in the Las Pilas experimental field, located at 19°46′48.85″ N, 104°13′57.65″ W [18].

Las Pilas is 860 m above sea level and has sandy loam soil with an average annual temperature of 23.9 °C and an average annual cumulative rainfall of 860 mm. Six commercial varieties were planted on the site (Table 1) of which the type of maturity they express has been identified; and 34 sugarcane hybrids of Mexican and foreign origin (Table 2), from the experimental fields of Instituto Nacional de Investigaciones Forestales Agrícolas y Pecuarias (INIFAP). The 34 hybrids were evaluated during the plant cycle and during the adaptability test phase.

The experimental unit was constituted of four furrows, each measuring 15 m in length and 1.40 m in width. The useful plot comprised the two central furrows,

Table 1 Commercial varieties and its type of maturity

ID	Commercial variety	Type of maturity in field
M6	CP 72-2086	Early
M3	ITV 92-1424	Early
M40	COLMEX 95-27	Early
M7	COLMEX 94-8	Early-Intermediate
M4	RB 85-5113	Early-Intermediate
M15	MEX 69-290	Intermediate-Late

Table 2 Materials established in Las Pilas experimental field

ID	Materials	Origin	ID	Materials	Origin
M1	COSTA JAL	Mexican	M23	ITSAMEX 06-6395	Mexican
M2	RB 85-5035	Abroad	M24	ITSAMEX 06-4863	Mexican
M5	LAICA 92-13	Abroad	M25	ITSAMEX 07-4954	Mexican
M8	ATEMEX 99-48	Mexican	M26	ITSAMEX 07-4387	Mexican
M9	ATEMEX 99-1	Mexican	M27	ITSAMEX 07-121115	Mexican
M10	ATEMEX 99-61	Mexican	M28	ITSAMEX 07-246	Mexican
M11	MEX 70-486	Mexican	M29	ITSAMEX 07-12116	Mexican
M12	TCP 89-3505	Abroad	M30	ITSAMEX 07-12113	Mexican
M13	MEX 80-1521	Mexican	M31	ITSAMEX 07-2963	Mexican
M14	COSTA JAL 2	Mexican	M32	ITSAMEX 07-99711	Mexican
M16	ITSAMEX 07-44814	Mexican	M33	ITSAMEX 07-9886	Mexican
M17	ITSAMEX 06-3049	Mexican	M34	ITSAMEX 07-86810	Mexican
M18	ITSAMEX 07-8681	Mexican	M35	ITSAMEX 07-12119	Mexican
M19	ITSAMEX 07-20810	Mexican	M36	ITSAMEX 07-1903	Mexican
M20	ITSAMEX 07-7501	Mexican	M37	ITSAMEX 07-44813	Mexican
M21	ITSAMEX 07-1107	Mexican	M38	ITSAMEX 07-12118	Mexican
M22	ITSAMEX 07-7259	Mexican	M39	CP 85-1382	Abroad

with the two outer meters excluded. All materials were established under irrigated conditions.

The 40 materials were planted on March 4, 2011, and 7 months after planting (October), the first measurements were taken. The measurement dates were October 8, 2011, November 8, 2011, December 8, 2011, and January 11, 2012. This period is referred to in the literature as the plant cycle, which typically lasts from 14 to 17 months. During this period, sugarcane undergoes four distinct stages: germination and/or emergence, tillering, rapid growth, and maturation.

The study included six industrial quality variables: sucrose, brix, purity, reductants, fiber, and humidity.

3 Latent Class Mixed Models

LCMMs represent a statistical approach that integrates elements of both LMMs and latent class models. LCMMs assume that the population is heterogeneous, comprising G latent classes of subjects, each characterized by a distinct mean trajectory profile. LCMMs are composed of three distinct components: a structural model, a measurement model, and a multinomial logistic model. The structural model explains the latent variable through covariates and time. The measurement model approximates the latent variable through observed variables. Finally, the multinomial logistic model describes the probability of each subject belonging to one of the latent classes [23].

For each subject i in a sample of N subjects, the structural model for subject i, $i = 1, \ldots, N$ at time t, is given by:

$$\Lambda_i(t) = \mathbf{X}_i(t)^\top \boldsymbol{\beta} + \mathbf{Z}_i(t)^\top \mathbf{b}_i + \boldsymbol{\omega}_i(t) \quad \forall t \in \mathbb{R}^+ \tag{1}$$

where

- $\Lambda_i(t)$ is a latent variable (latent process);
- $\mathbf{X}_i(t)$ is a vector of p covariates at time t associated with the vector of fixed effects $\boldsymbol{\beta}$;
- $\mathbf{Z}_i(t)$ is a vector of q covariates at time t associated with the vector of random effects $\mathbf{b}_i$;
- $\mathbf{b}_i$ follows a multivariate normal distribution with mean vector $\mathbf{0}$ and variance–covariance matrix $\mathbf{D}$, i.e., $\mathbf{b}_i \sim \mathcal{N}_{n_i}(\mathbf{0}, \mathbf{D})$.
- $\boldsymbol{\omega}_i(t)$ is an error term associated with subject i at time t, capturing the variability not explained by the observed and unobserved covariates. It is assumed that $\boldsymbol{\omega}_i(t) \sim \mathcal{N}_{n_i}(\mathbf{0}, \sigma_\omega^2 \boldsymbol{\Omega}_i)$.

For each subject i, $i = 1, \ldots, N$, there are n_i measurements over time, $j = 1, \ldots, n_i$; thus, $Y_i = (Y_{i1}, \ldots, Y_{ij}, \ldots, Y_{in_i})^\top$ represents the vector of observations for subject i. The measurement model is defined by the relationship between the latent process $\Lambda_i(t)$ and the observed value Y_i through the link function $H(\cdot)$ [24, 25].

$$Y_i = H\left(\Lambda_i\left(t\right) + \boldsymbol{\epsilon}_i; \boldsymbol{\eta}\right) \tag{2}$$

where

- $\boldsymbol{\epsilon}_i$ represents independent Gaussian errors with a variance σ_ϵ^2, in other words, $\boldsymbol{\epsilon}_i \sim \mathcal{N}_{n_i}(\mathbf{0}, \sigma_\epsilon^2 \mathbf{I}_{n_i})$. This term represents the variability that is not explained by the latent process.
- $\boldsymbol{\eta}$ is the vector of parameters associated with the link function $H(\cdot)$.

The multinomial logistic model is defined as follows:

$$\pi_{ig} = P\left(c_i = g \mid \mathbf{X}_{ci}\right) = \frac{e^{\xi_{0g} + \mathbf{X}_{ci}^{\top} \boldsymbol{\xi}_{1g}}}{\sum_{k=1}^{G} e^{\xi_{0k} + \mathbf{X}_{ci}^{\top} \boldsymbol{\xi}_{1k}}}, \; g = 1, \ldots, G \tag{3}$$

where

- π_{ig} is the conditional probability (posterior probabilities) that the latent class c_i is equal to g given the set of covariates $\mathbf{X}_{ci}$;
- ξ_{0g} is the intercept of the class g; for identifiability, it is assumed $\xi_{0G} = 0$;
- $\boldsymbol{\xi}_{1g}$ is a vector of coefficients for the latent class g associated with the vector of covariates $\mathbf{X}_{ci}$, by identifiability, it is assumed $\xi_{1G} = 0$;
- In the denominator, it is summed over all latent classes.

To analyze the genotypes of *Saccharum spp*, we have obtained four repeated measurements over time. Therefore,

$$\mathbf{X}_i = \mathbf{Z}_i = \begin{pmatrix} 1 & t_1 \\ 1 & t_2 \\ 1 & t_3 \\ 1 & t_4 \end{pmatrix}, \; \boldsymbol{\beta} = \begin{pmatrix} \beta_0 \\ \beta_1 \end{pmatrix}, \; \mathbf{b}_i = \begin{pmatrix} b_{0i} \\ b_{1i} \end{pmatrix}, \; \boldsymbol{\epsilon}_i = \begin{pmatrix} \epsilon_{i1} \\ \epsilon_{i2} \\ \epsilon_{i3} \\ \epsilon_{i4} \end{pmatrix}$$

then

$$\Lambda_i(t) = \mathbf{X}_i(t)^{\top} \boldsymbol{\beta} + \mathbf{Z}_i(t)^{\top} \mathbf{b}_i, \; i = 1, \ldots, 40. \tag{4}$$

The variables studied were measured in months. In the structural model for the materials, no correlation structure was assumed for $\boldsymbol{\omega}_i(t)$, $i = 1, \ldots, 40$.

The measurement model for subject i and class g is as follows:

$$Y_i \mid_{c_i = g} = \mathbf{X}_{1i}(t)^{\top} \boldsymbol{\beta} + \mathbf{X}_{2i}(t)^{\top} \boldsymbol{\alpha}_g + \mathbf{Z}_i(t)^{\top} \mathbf{b}_{ig} + \boldsymbol{\epsilon}_i \tag{5}$$

where $\mathbf{X}_{1i}$ represents the matrix of covariates for fixed effects for all classes. Similarly, $\mathbf{X}_{2i}$ denotes the matrix of covariates for fixed effects for each class. Finally, $\mathbf{Z}_i$ is the matrix of covariates associated with the random effects of subject i in class g. In parallel, the structural model for subject i and class g is [24]:

$$\Lambda_i(t) \mid_{c_i = g} = \mathbf{X}_{1i}(t_i)^{\top} \boldsymbol{\beta} + \mathbf{X}_{2i}(t_i)^{\top} \boldsymbol{\alpha}_g + \mathbf{Z}_i(t_i)^{\top} \mathbf{b}_{ig} \tag{6}$$

For the subject i, $i = 1, \ldots, N$, the matrix $\mathbf{X}_{ci}$ in the multinomial logistic model is defined as follows:

$$\mathbf{X}_{ci} = \begin{pmatrix} 1 & t_1 \\ 1 & t_2 \\ 1 & t_3 \\ 1 & t_4 \end{pmatrix}$$

The multinomial logistic model determines the probability of belonging to the latent class g, $g = 1, ..., G$ for each of the $N = 40$ materials.

3.1 Statistical Software

For statistical analysis, we used R [26]. The mixed latent class model was fitted using the *hlme* function from the *lcmm* package [27].

The parameters of the LCMM were estimated using the maximum likelihood method.

The fit of the model was determined using the AIC criterion, which assesses the fit of the model to the data and the complexity of the model [28].

4 Results

Table 3 shows the mean values of the four time points for each of the 40 materials, for the six industrial quality variables. For the six industrial quality variables, the means of the six commercial varieties are similar to the means of the 34 genotypes studied.

An LCMM was fitted to each of the six industrial quality variables. Although sucrose percentage is the most commonly used variable to evaluate new hybrids in the adaptability testing phase, this variable did not show significant differences in the grouping of materials with the LCMM. Only the industrial quality variable fiber percentage showed significant differences in the grouping of the 40 materials (see Figs. 3 and 4). For this variable, the LCMMs reported the smallest AIC for two latent classes, $g = 2$, grouping 30% of the materials in the first class and 70% of the materials in the second class (Table 4).

Table 5 shows the estimation of the fixed effects parameters for the two latent classes. According to the p-values of the Wald statistic, all parameters in the model are significant, indicating that the materials can be grouped into two classes. The value of the intercept for Class 1 (6.13) represents the initial level or expected value of the response variable (fiber percentage) at the beginning of the study. Meanwhile, time (0.44) has a positive effect, indicating the expected change in the response variable per unit of time for this class. For the latent Class 2, the intercept value is 3.21. Similar to the previous case, this value represents the initial value of the fiber variable for this class. However, time has a higher positive effect, indicating a faster change in the response variable per unit of time. The values of these parameters indicate how the fiber variable will evolve.

Numerically, Class 1 has a higher intercept than Class 2. This means that materials in Class 1 have a higher percentage of fiber in the first measurement than materials in Class 2. On the other hand, the estimated slope of the time for Class 1

Table 3 Averages of the 40 materials in the industrial quality variables

ID	Materials	Sucrose	Brix	Purity	Reducing agents	Fiber	Moisture
M1	COSTA JAL	10.845	14.537	73.920	0.422	11.867	74.150
M2	RB 85-5035	10.308	14.005	73.425	0.577	11.705	73.922
M3	ITV 92-1424	11.810	15.110	77.868	0.604	11.378	73.403
M4	RB 85-5113	11.683	15.393	75.828	0.623	11.447	74.703
M5	LAICA 92-13	9.842	13.700	71.493	0.628	11.570	76.320
M6	CP 72-2086	11.447	14.978	76.078	0.539	11.445	75.250
M7	COLMEX 94-8	12.895	16.038	79.998	0.517	12.085	72.585
M8	ATEMEX 99-48	12.870	16.130	79.695	0.320	11.992	73.153
M9	ATEMEX 99-1	10.148	14.092	71.995	0.679	9.773	77.765
M10	ATEMEX 99-61	12.100	15.643	77.287	0.415	11.407	73.963
M11	MEX 70-486	11.877	15.412	76.545	0.570	10.855	75.472
M12	TCP 89-3505	7.788	12.160	63.010	0.892	10.957	77.877
M13	MEX 80-1521	10.725	14.550	73.590	0.469	10.732	76.558
M14	COSTA JAL 2	8.125	12.440	62.013	0.894	10.655	79.400
M15	MEX 69-290	10.845	14.600	72.040	0.707	9.982	77.440
M16	ITSAMEX 07-44814	10.000	13.940	71.600	0.628	11.305	75.120
M17	ITSAMEX 06-3049	10.023	13.950	70.740	0.741	11.412	74.403
M18	ITSAMEX 07-8681	9.985	13.508	72.940	0.885	10.460	76.600
M19	ITSAMEX 07-20810	7.420	12.060	61.233	1.002	10.790	77.980
M20	ITSAMEX 07-7501	8.393	12.498	71.737	0.977	9.818	78.820
M21	ITSAMEX 07-1107	8.408	12.560	66.525	1.017	11.863	75.810
M22	ITSAMEX 07-7259	9.178	13.027	68.113	1.032	10.967	76.025
M23	ITSAMEX 06-6395	11.895	15.260	77.850	0.575	10.473	75.412
M24	ITSAMEX 06-4863	10.008	13.645	74.218	0.782	9.957	77.540
M25	ITSAMEX 07-4954	7.237	11.502	62.715	1.062	11.018	77.865
M26	ITSAMEX 07-4387	8.768	12.678	67.617	0.990	9.395	78.677
M27	ITSAMEX 07-121115	8.110	12.242	62.358	1.128	9.703	77.407
M28	ITSAMEX 07-246	8.530	12.545	65.683	1.022	11.110	78.190
M29	ITSAMEX 07-12116	9.393	13.402	69.245	0.717	11.685	76.408
M30	ITSAMEX 07-12113	8.890	12.947	67.030	0.947	11.373	75.944
M31	ITSAMEX 07-2963	9.445	13.560	69.127	0.963	10.695	77.463
M32	ITSAMEX 07-99711	9.742	13.562	71.428	0.902	9.850	77.303
M33	ITSAMEX 07-9886	10.068	13.707	72.858	0.869	11.658	74.985
M34	ITSAMEX 07-86810	8.700	12.742	66.925	0.607	10.610	78.445
M35	ITSAMEX 07-12119	9.598	13.877	72.075	1.121	10.475	77.082
M36	ITSAMEX 07-1903	10.155	13.735	73.625	0.826	9.607	77.622
M37	ITSAMEX 07-44813	11.408	14.795	76.278	0.936	11.225	75.453
M38	ITSAMEX 07-12118	10.680	14.288	72.900	1.147	10.168	76.753
M39	CP 85-1382	12.680	15.730	80.470	0.862	11.370	74.213
M40	COLMEX 95-27	11.060	14.575	75.483	0.727	10.870	76.417

Table 4 Statistics and classification by material group for the fiber variable with LCMM

Groups	AIC	% Class 1	% Class 2	% Class 3	% Class 4	% Class 5
1	473.03	100.00				
2	471.18	30.00	70.00			
3	478.43	32.50	62.50	5.00		
4	486.43	0.00	32.50	62.50	5.00	
5	494.43	5.00	0.00	62.50	0.00	32.50

Table 5 Fixed effects estimation of the multinomial logistic model (3)

Parameters	Coefficient	Standard error	Wald test	p-value
Intercept Class 1	6.13	1.02	6.00	< 0.01
Time Class 1	0.44	0.11	3.77	< 0.01
Intercept Class 2	3.21	0.66	4.83	< 0.01
Time Class 2	0.92	0.07	12.03	< 0.01

is lower than the Class 2, meaning that materials in Class 1 do not accumulate as much fiber percentage over time as class 2 materials.

In Fig. 1, the four measurements of the percentage of fiber for the 40 materials are shown. Graphically, approximately constant and increasing trends can be observed. The six commercial varieties MEX 69-290 (M15), COLMEX 95-27 (M40), ITV 92-1424 (M3), CP 72-2086 (M6), COLMEX 94-8 (M7), and RB 85-5113 (M4) exhibited an increasing trend, indicating that their percentage of fiber increases over time.

Among the 34 new genotypes, only in the genotypes ITSAMEX 07-1903 (M36), ITSAMEX 07-121115 (M27), and ATEMEX 99-1 (M9) no increase in fiber percentage was observed. This could be due to factors such as changes in growing conditions, sugarcane maturation, or other environmental variables.

The number and percentage of materials classified as Class 1 and Class 2 are shown in Table 6; 30% of the materials were classified as Class 1, while 70% were classified as Class 2. The conditional probabilities for Class 1 and Class 2 are 0.91 and 0.96, respectively, indicating that 91% of the materials are in Class 1 and 96% are in Class 2. The conditional probabilities for three thresholds are also shown. 91.67% of the materials are classified in Class 1 with a conditional probability above 0.7, while 92.86% of the materials are classified in Class 2 with a conditional probability above 0.7. For a threshold of 0.8, 83.33% of the materials are classified in Class 1, and 92.86% of the materials are classified in Class 2. Finally, 75% of the materials are classified in Class 1 with a conditional probability above 0.9, while 89.29% of the materials are classified in Class 2 with a conditional probability above 0.9. These results demonstrate the clustering of materials based on the conditional probabilities obtained from the multinomial logistic model.

Figure 2 shows the distribution of the 40 sugarcane materials in their respective latent classes. This graph provides information on how the materials cluster and differentiate based on the conditional probabilities obtained from the multi-

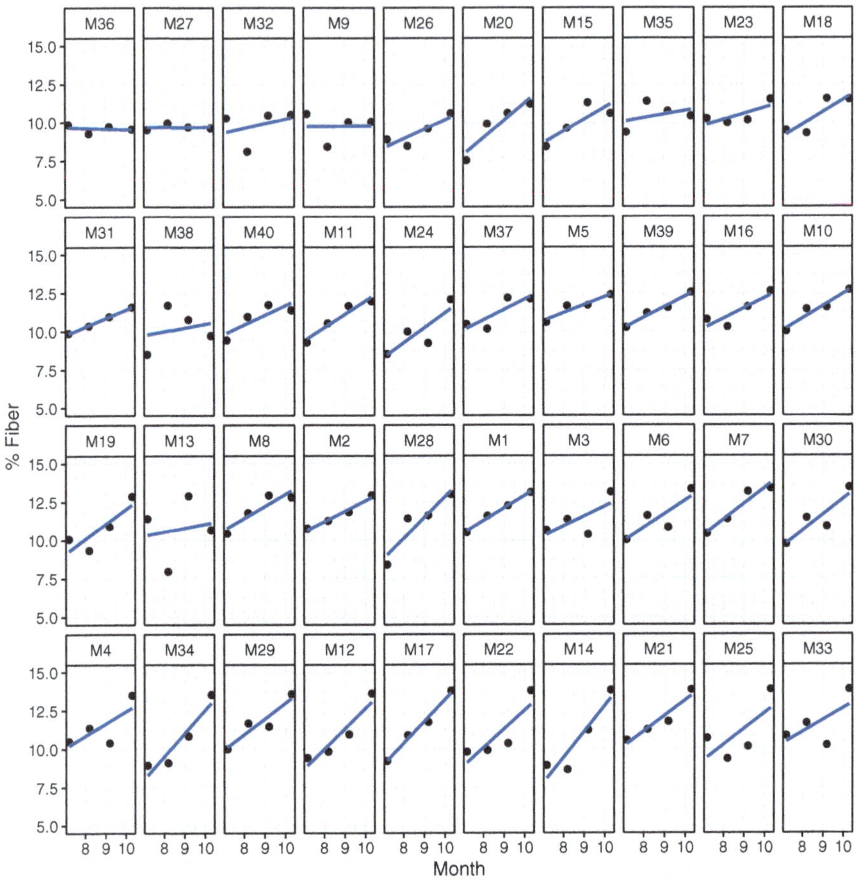

Fig. 1 Fiber percentage observed in the 40 materials

Table 6 Posterior classification and posterior probabilities above a threshold (%)

Result	Class 1	Class 2
N	12	28
%	30	70
Prob1	0.91	0.04
Prob2	0.09	0.96
Prob > 0.7	91.67	92.86
Prob > 0.8	83.33	92.86
Prob > 0.9	75.00	89.29

nomial logistic model using latent class mixed models. Materials above 0.75 have a high probability of belonging to a particular class. Only the materials ITSAMEX 07-8681 and MEX 80-1521 have a conditional probability close but greater than 0.50, indicating a less defined classification for these materials. In

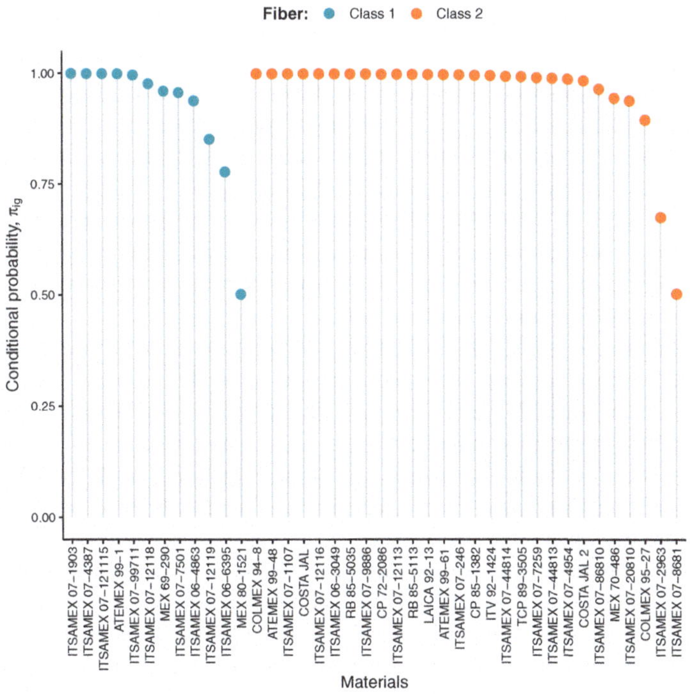

Fig. 2 Clustering of the 40 materials by percentage of fiber

Class 1, the commercial variety MEX 69-290 (M15) is found together with the materials ATEMEX 99-1 (M9), MEX 80-1521 (M13), ITSAMEX 07-7501 (M20), ITSAMEX 06-6395 (M23), ITSAMEX 06-4863 (M24), ITSAMEX 07-4387 (M26), ITSAMEX 07-121115 (M27), ITSAMEX 07-99711 (M32), ITSAMEX 07-12119 (M35), ITSAMEX 07-1903 (M36) and ITSAMEX 07-12118 (M38). The remaining materials are in Class 2, including the commercial grades CP 72-2086 (M6), ITV 92-1424 (M3), COLMEX 95-27 (M40), COLMEX 94-8 (M7), and RB 85-5113 (M4).

In Class 1, the materials with a lower percentage of fiber are found, and in Class 2, the materials with a higher percentage of fiber. It can also be observed that in Class 1 the commercial variety with medium-late maturity is found, while in Class 2 the commercial varieties with early and early-intermediate maturity are present. In this sense, the grouping of materials has an agronomic meaning, because the hybrids grouped in Class 1 are of medium-late maturity, while the hybrids in Class 2 are of early and early-intermediate maturity.

5 Discussion and Conclusion

The completion of this work allowed the application of a statistical technique, rarely used in agricultural sciences, to a problem of importance in the sugar industry. This application represents a step forward since most of the work in this field uses techniques for cross-sectional data, such as analysis of variance or polynomial regression models. The LCMMs allowed the analysis of the six industrial quality variables from a temporal perspective, that is, studying the variation of each of these variables during the plant cycle period (germination and/or emergence, tillering, rapid growth, and maturation) and not only in a single stage.

According to the results, the fitted latent class mixed model showed differences in the rate of fiber between the materials studied in the crop cycle adaptability test phase. In the variable, the percentage of sucrose, one of the most commonly used variables to identify materials with higher performance in the field, no differences were identified between the 34 new materials and the commercial varieties evaluated. Similarly, no differences were found in the brix, purity, reducers, and moisture variables.

For future work, environmental variables can be included in the analysis. The flexibility of the LCMM is reflected in the linear prediction of fixed effects, as nonlinear covariates that affect the response variable can be included. Another factor contributing to the flexibility of these models is the ability to choose, based on experience or exploratory analysis, the covariates that most affect variability. The ability to decide which factors affect variability is rare in most statistical models because they typically assume a defined variability for the model.

This work sets a precedent for studying the performance of sugarcane materials in the early selection stages.

Appendix

The following plots show the behavior of the residuals. A good fit is indicated when the residuals have an approximate normal distribution with a mean of 0 and a standard deviation of 1. The figure shows that the marginal residuals fall within the range from -2 to 2, similar to the specific residuals of each genotype. This indicates a good goodness of fit when random effects are included. We also observe the quantile–quantile plot, where the marginal residuals adhere to the confidence bands until they reach higher quantiles. However, the specific residuals of each genotype adhere closely to the confidence bands and deviate only from 0 at the quantiles further away (Figs. 3 and 4).

Fig. 3 AIC for the percentage of fiber

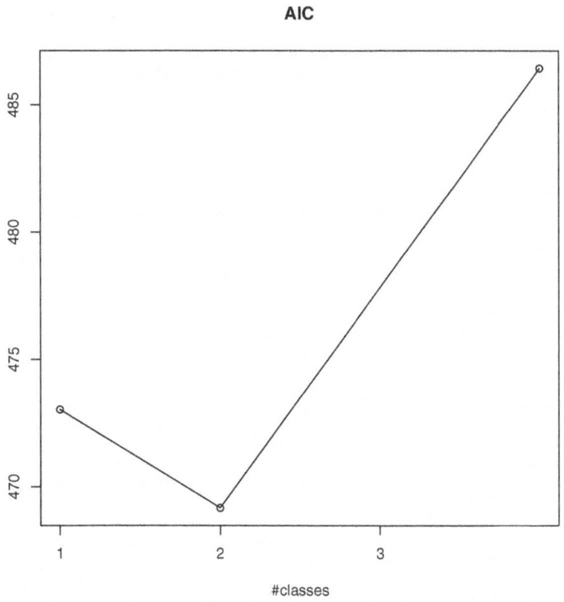

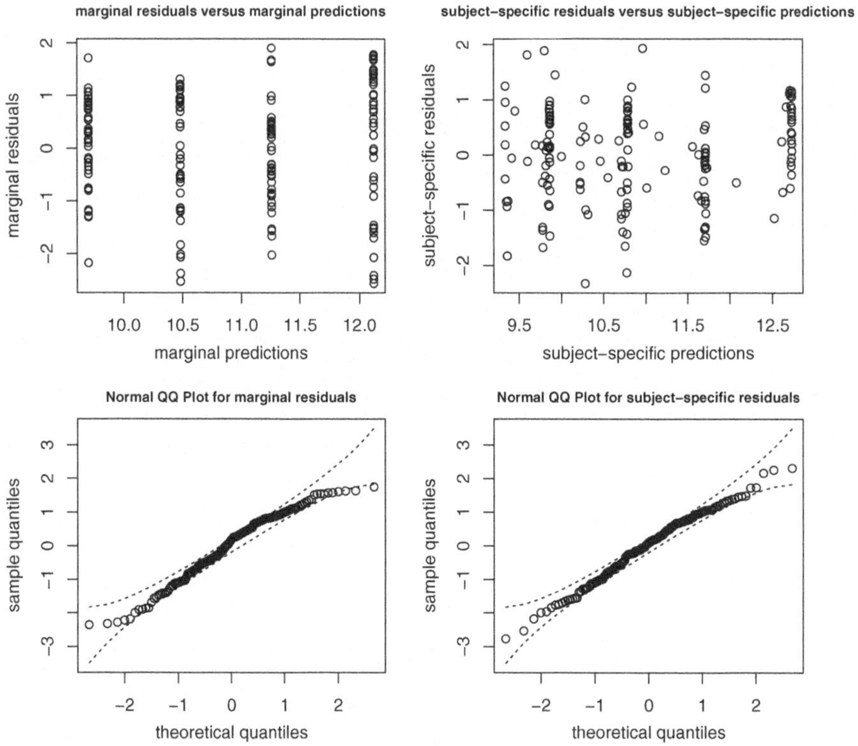

Fig. 4 Subject-specific and marginal residuals in the two-class linear mixed model

Acknowledgments The Technical Field Department of Ingenio Melchor Ocampo S.A. de C.V., Engineer Filemón Zavalza García, and the Cane Production and Quality Committee of the said mill. Special thanks to Gabriel Ricardo Blackaller Ayala and Jesús Zúniga Mendoza.

References

1. Ilich Miguel Santiago Zárate et al. "Productividad y cambio tecnológico en la agroindustria de la caña de azúcar en México." In: *Revista mexicana de ciencias agrícolas*, 12.6 (2021), pp. 1005–1017, 2021.
2. Braulio Villatoro. *Uso de índices de vegetación NDWI y MSI obtenidos de imágenes satelitales Sentinel-2 y Landsat 8 para monitoreo de humedad, como indicador de momento óptimo de cosecha en caña de azúcar*, 2021.
3. JE Larrahondo et al. "El cultivo de la caña de azúcar en la zona azucarera de Colombia". In: *En: Cassalett, C.. Torres, J. y Isaacs, C., Eds., Calidad de caña. Cenicaña, Cali*, (1995) pp. 337–354.
4. Yulexi Mendoza Batista et al. "Comportamiento de los cultivares de caña de azúcar C97-445 y C95-416 en localidades de la provincia Holguín. *Centro Agrícola*, 46.1 (2019), pp. 49–53.
5. Irenaldo Delgado Mora et al. "Influencia de la edad y cultivar de caña de azúcar en el momento de la cosecha." In: *Centro Agrícola*, 43.2 (2016), pp. 59–65.
6. H Ramírez-Cathí et al. "Rendimiento, caracterización morfológica y bromatológica de la punta de caña de azúcar en la Huasteca Potosina, México. "In: *Revista Cubana de Ciencia Agrícola*, 48.4 (2014), pp. 411–415.
7. Jesús E Larrahondo. "Calidad de la caña de azúcar." In: *El cultivo de la caña en la zona azucarera de Colombia. Eds. Cassalett, C*, (1995), pp. 337–354.
8. CONADESUCA. *Nuevas variedades de caña de azúcar*, 2016.
9. IMPA. *Informe técnico 1989, córdova, ver. mex.*, 1990.
10. Robert A Gilbert et al. "Sucrose accumulation and harvest schedule recommendations for CP sugarcane cultivars". In: *Crop Management*, 3.1 (2004) pp. 1–7.
11. José Reyes-Hernández et al. "Rendimiento y calidad de siete variedades de caña de azúcar en EL Mante, Tamaulipas". In: *Revista mexicana de ciencias agrícolas*, 13.5 (2002), pp. 883–892,
12. F. Martín-Armas. *Estudio de adaptabilidad al genotipo ambiente en las variables agroindustriales de siete variedades de caña de azúcar de origen cubano, en diferentes tipos de cepas, desde los 400 a 1100 msnm., en la provincia de orona Santiago*, 2009.
13. Reynaldo Rodríguez et al. "Estudio del rendimiento y modelación del período de madurez en nuevos cultivares de caña de azúcar." In: *Cultivos Tropicales*, 36.4 (2015), 134–143.
14. Shine J.M. Miller J.D. Rice R.W. & Odero D.C. Gilbert, R.A. "Maturity Crves and Harvest Schedule Recommendations for CP Sugarcane Varieties1". In: (2004).
15. Shine J.M. Miller J.D. & Rice R.W. Gilbert, R.A. "Sucrose accumulation and harvest schedule recommendations for CP sugarcane cultivars". *Crop Management*, 3.1 (2004), pp.0–0.
16. T.D.R. Martínez. "*Evaluación agroindustrial de variedades de caña de azúcar (Saccharum officinarum L.) en el Ingenio El Modelo, Úrsulo Galván, Veracruz. Campus Veracruz, Colegio de Postgraduados. Veracruz, México*. PhD thesis, Tesis de Maestría, 2009.
17. & Sousa Vieira O.D. Rea, R. "Genotype x environment interactions in sugarcane yield trials in the central-western region of Venezuela". In: *Interciencia*, 27.11 (2002).
18. JC García-Preciado. "Evaluación de variables de calidad en híbridos de saccharum spp en diferentes ambientes agroecológicos de Jalisco, México." In: *Agro Productividad*, 10.11 (2017).
19. Santiago Ostengo et al. "A protocol for identifying characteristic sucrose accumulation curves of sugarcane genotypes (Saccharum spp.)." In: *Sugar Tech*, 23 (2021), pp. 519–523, DOI: https://doi.org/10.1007/s12355-020-00926-8.

20. Emmanuel Bizimungu and Nassul Kabunga. "A latent class analysis of improved agro-technology use behavior in Uganda: Implications for optimal targeting." In: (2018).
21. Gregory Howard and Brian E Roe. "Stripping because you want to versus stripping because the money is good: a latent class analysis of farmer preferences regarding filter strip programs." In: (2013).
22. Shawn P Gilroy et al. Beyond Systematic and Unsystematic Responding: Latent Class Mixture Models to Characterize Response Patterns in Discounting Research." In: *Frontiers in behavioral neuroscience*, 16 (2022), p. 806944.
23. Carlomagno Araya Alpízar. "Diagnóstico de modelos de clases latentes en tablas con frecuencias pequeñas o nulas: efecto de sumar a las frecuencias de los patrones de respuestas." In: *XII Congreso Galego de Estatística e Investigación de Operacións: Lugo, 22-23-24 de outubro de 2015: actas*. Servizo de Publicacións. 2015, pp. 48–50.
24. Cécile Proust-Lima, Viviane Philipps, and Benoit Liquet. "Estimation of extended mixed models using latent classes and latent processes: the R package lcmm." In: *arXiv preprint arXiv:1503.00890* (2015).
25. Roberto Burro et al. "An estimation of a nonlinear dynamic process using latent class extended mixed models: affect profiles after terrorist attacks. In: *Nonlinear Dyn. Psychol. Life Sci*, 22.1 (2018) pp. 35–52
26. R Core Team. *R: A Language and Environment for Statistical Computing*. R Foundation for Statistical Computing, Vienna, Austria, 2023. URL: https://www/R-project/org/.
27. Cecile Proust-Lima et al. "Package 'lcmm'. In: *CRAN R*, (2023).
28. Sandra L Posada and Ricardo Rosero Noguera. "Comparación de modelos matemáticos: una aplicación en la evaluación de alimentos para animales". In: *Revista Colombiana de Ciencias Pecuarias* 20.2 (2007), pp. 141–148.

Kernel Feature Ordering by Conditional Independence in Species Diversity

Gabriel A. Rodríguez-Yam and José P. Sánchez-Venado

Abstract The *Kernel Feature Ordering by Conditional Independence* (KFOCI) algorithm (Huang et al. J Mach Learn Res 23(216):1–58;2022. http://jmlr.org/papers/v23/21-493.html) is a variable selection technique that does not assume a functional form for the model. In the present study, the KFOCI algorithm was implemented to search the variables that explain the spatial variation patterns of species diversity in a mountain system analyzed in Martínez-Camilo et al. (Biotropica 50(3):499–509;2018). We managed to obtain a subset of covariates that has smaller *Bayesian Information Criterion* (BIC) value when we use the parametric model implemented in Martínez-Camilo et al. (Biotropica 50(3):499–509;2018) than that obtained for the subset of covariates obtained in Martínez-Camilo et al. (Biotropica 50(3):499–509;2018). Both subsets have only climate variables, suggesting that climate variables are more important than the other factors in explaining diversity variation in mountain systems.

Keywords Kernel Feature Ordering by Conditional Independence · Kernel partial correlation coefficient · Model-free nonlinear variable selection · Nearest neighbor methods · Bayesian Information Criterion

1 Introduction

An important goal in statistical modeling is to select the "best" subset of covariates in a given model. There are many selection algorithms for parametric models that are consistent when the assumptions are true, but their performance may decrease as the data generating process becomes more complicated. For this reason, algorithms that do not assume a functional form, known as nonparametric methods,

G. A. Rodríguez-Yam (✉) · J. P. Sánchez-Venado
Department of Statistics, Mathematics and Computing, Chapingo Autonomous University, Texcoco, Mexico
e-mail: grodriguezy@chapingo.mx

J. A. Christen et al. (eds.), *Statistics, Society and Environment*, Springer Proceedings in Mathematics & Statistics 479, https://doi.org/10.1007/978-3-031-78401-9_13

are gaining importance. These are good techniques when interactions are present, but they depend on goodness of fit, and their theoretical foundations are often weak [3]. However, this can be overcome in [3] with the *Kernel Feature Ordering by Conditional Independence* (KFOCI), a consistent, model-free variable selection method that, under a set of assumptions, uses a conditional dependence measure to select the subset of covariates whose complement is independent of the response variable. Following this line of thought, a new measure of conditional dependence between random variables that can take values in general topological spaces is proposed in [12]. They succeeded in generalizing the method in [3] with strong theoretical properties and the possibility of including categorical and even non-Euclidean variables.

There are many complex natural phenomena that are difficult to adequately describe with parametric models [10]. An example is the study of spatial variation patterns of diversity in mountain systems. Although these mechanisms remain controversial [2, 9], it is known that climatic factors such as rain, temperature, or the combination of these are involved [14, 17]. However, the inconsistencies shown in some studies [21, 22] suggest that the interactions between these factors and many others are complex [4, 7]. The main objective of this study is to illustrate that KFOCI can be extremely helpful in studies like these. For this purpose, KFOCI was implemented to the dataset from the study of the mountain system in [16] to select a "good" subset of covariates from this dataset without a parametric model. To compare the subsets obtained by KFOCI with the subset of covariates selected parametrically in [16], the *Bayesian Information Criterion* (BIC) [19], a measure of trade-off between goodness of fit and the number of parameters of the parametric model, was considered. This criterion "encourages" the selection of simpler models and helps prevent overfitting [5], as it often happens with the *Akaike Information Criterion* [1].

In the next section, the KFOCI algorithm will be briefly explained. In the third section, the KFOCI algorithm will be implemented to the database analyzed in [16], and the results will be compared with those obtained in [16]. Finally, in the last section, some conclusions will be provided.

2 Kernel Feature Ordering by Conditional Independence

The algorithm *Kernel Feature Ordering by Conditional Independence* (KFOCI) proposed in [12] is based in the *Kernel Partial Correlation* (KPC) coefficient given by

$$\rho^2(Y, Z|X) := \frac{\mathbb{E}[\mathrm{MMD}^2(P_{Y|XZ}, P_{Y|X})]}{\mathbb{E}[\mathrm{MMD}^2(\delta_Y, P_{Y|X})]},$$

where $(X, Y, Z) \sim P$ is a random element on the completion of a product of appropriate topological spaces, MMD is the *Maximum Mean Discrepancy* [8], and

δ_Y is the Dirac measure. This coefficient is interpreted as a generalization of the classical partial correlation that captures the conditional association between two random variables.

Let us consider a *general regression model* $Y = f(X_1, \ldots, X_p, \epsilon)$, where Y is a response variable on the topological space $\mathcal{Y}$, $X_1, \ldots, X_p$ are covariates on metric spaces, f is an unknown function, and ϵ is an unobserved error independent of $X_1, \ldots, X_p$. Let $X_S := (X_j : j \in S)$ for $S \subseteq \{1, \ldots, p\}$. S is called a *sufficient subset* if $Y \perp\!\!\!\perp X_{S^c} \mid X_S$. In [12], the problem of *variable selection* is to select a sufficient subset S, ideally with the smallest cardinality, such that $Y = \mathbb{E}[f(X_1, \ldots, X_n, \epsilon)|X_S, \epsilon] =: g(X_S, \epsilon)$, where g is the unknown function of the "updated" general regression of Y against X_S. The main idea of the KFOCI stepwise algorithm is as follows. If the covariates $X_{j_1}, \ldots, X_{j_k}$ have already been selected, they propose to find $X_{j_{k+1}}$ that maximizes $\rho^2\left(Y, X_{j_{k+1}}|X_{\{j_1,\ldots,j_k\}}\right)$. But they show that this is equivalent to finding $X_{j_{k+1}}$ that maximizes $T(\{j_1, \ldots, j_{k+1}\})$ given by $T(S) := \mathbb{E}\{\mathbb{E}[k_{\mathcal{Y}}(Y, Y')|X_S]\}$, where $k_{\mathcal{Y}} : \mathcal{Y} \times \mathcal{Y} \to \mathbb{R}$ is the kernel of the respective separable *reproducing kernel Hilbert space* over $\mathcal{Y}$ associated with the MMD, and $Y, Y' \overset{iid}{\sim} P_{Y|X_S}$.

If $\{(Y_i, X_{1i}, \ldots, X_{pi}) : i = 1, \ldots, n\}$ is a random sample from the joint distribution of $(Y, X_1, \ldots, X_p)$, then a consistent estimator of $T(S)$ is the following

$$T_n(S) := \frac{1}{n}\sum_{i=1}^{n}\frac{1}{d_i}\sum_{i':(i,i')\in\mathcal{E}(\mathcal{G}_n)} k_{\mathcal{Y}}(Y_i, Y_{i'}),$$

where if $X_{S,i} := \{X_{ji} : j \in S\}$, then $\mathcal{G}_n := \mathcal{G}(X_{S,1}, \ldots, X_{S,n})$ and $\mathcal{G}$ is some *geometric graph functional*, such that two vertices are connected by an edge in $\mathcal{G}_n$ if they are "close." $\mathcal{E}(\mathcal{G}_n)$ is the set of edges of $\mathcal{G}_n$, and d_i is the degree of $X_{S,i}$ in $\mathcal{G}_n$. The *K-nearest neighbor graph*, which does not have any multiple edges and self loops, is selected. The KFOCI algorithm to find $\hat{S}$ is the following:

Initialization $k = -1,\ \hat{S} = \emptyset,\ \{j_0\} = \emptyset$, and $T_n(\emptyset) = -\infty$
Do

1. $k = k + 1$
2. $\hat{S} = \hat{S} \cup \{j_k\}$
3. $j_{k+1} := \arg \max_{\ell\in\{1,\ldots,p\}\setminus\hat{S}} T_n(\{j_1, \ldots, j_k, \ell\})$

while $T_n(\hat{S} \cup \{j_{k+1}\}) \geq T_n(\hat{S})$ and $k < p$.

3 Species Diversity

In this section, the KFOCI algorithm was implemented to the database analyzed in [16]. This dataset comes from an experiment that aims to search for the mechanisms

Table 1 Explanatory variables studied in [16]. The first column shows identifiers that will be used in Table 2

Variable	Description	Variable	Description
amt	Annual mean temperature	*nitro*	Nitrogen
ar	Annual rainfall	*om*	Organic matter
cec	Cation exchange capacity	*phos*	Phosphorus
clay	Clay content	*ph*	Potential of hydrogen
elev	Elevation	*rcq*	Rainfall of coldest quarter
isot	Isothermality	*rdm*	Rainfall of driest month
lati	Latitude	*rdq*	Rainfall of driest quarter
long	Longitude	*rwq*	Rainfall of warmest quarter
mtwm	Max temperature of warmest month	*rwm*	Rainfall of wettest month
mdr	Mean diurnal range	*rweq*	Rainfall of wettest quarter
mtcq	Mean temperature of coldest quarter	*rs*	Rainfall seasonality
mtdq	Mean temperature of driest quarter	*slope*	Slope
mtwq	Mean temperature of warmest quarter	*tar*	Temperature annual range
mtweq	Mean temperature of wettest quarter	*ts*	Temperature seasonality
mtcm	Min temperature of coldest month	*tran*	Transect

that explain the patterns of spatial variation of diversity in mountain systems, an ongoing topic [2, 9, 18]. The sample size in this dataset was of 128. It contains three response variables and 30 explanatory variables that are listed in Table 1. The explanatory variables operate on local and regional levels, one of which, the elevation transect, is qualitative (categorical).

According to [16], it is argued that it is convenient to focus on tree species only, dividing them into two functional groups: pioneer species and late-successional species. The response variable, the *tree species diversity*, was calculated using Shannon's exponential index, evaluated in the two functional groups and in the total species (both groups). Our study will focus only on the total species.

In [16], the response variable was modeled with a GLM using a log link and the gamma distribution for the response variable. Their variable selection method does not consider any classical information criterion, but two phases of variable selection. In the first phase, they eliminate collinear variables. In this phase, they selected nine covariates, namely cec, $clay$, $elev$, $isot$, $phos$, rdm, rwq, $slope$, and tar. In the second phase, they use the statistical significance of the GLM coefficients. Their final model contained the two covariates shown in Table 2 as subset H, namely $isot$ and rwq.

To implement the KFOCI algorithm, we used the `KFOCI` function from the **KPC** package developed in the R-project [11]. The kernel $k_{\mathcal{Y}}(y, y') = \exp\left(\frac{\|y-y'\|^2}{2s^2}\right)$, where s is the median of the pairwise distances $\{\|y_i - y_j\|\}_{i<j}$ was used. To avoid ties, we considered an odd number of neighbors (K), namely $K = 1, 3, 5, 7, 9$ [15, p. 65]. KFOCI is a stochastic algorithm, i.e., if it is run twice, the selected subset may differ each another; therefore, it was run 100 times for each value of K. The results are summarized in Table 2.

Table 2 KFOCI results. Columns 4 and 5 display the sufficient subsets and their frequencies for each value of K. An identifier for the sufficient set is shown in column 3. In the last column, the BIC values based on the GLM model in [16] are shown

Algorithm	K	Subset	Selected variables	Frequency	BIC
KFOCI	1	A	*rs, rdq, rcq*	84%	793.82
		–	Another 5 subsets	16%	–
	3	B	*rs, rwq*	100%	806.61
	5	C	*tran, elev, isot, rwm, rwq, lati*	99%	–
		C'	*tran, tar*	1%	–
	7	D	*lati, long*	97%	806.15
		D'	*tran, tar*	3%	–
	9	E	*tran, ph, tar, rs, rwq*	86%	–
		–	Another 2 subsets	14%	–
Proposed in [16]	–	F	*isot, rwq*	–	803.01
GLM stepwise	–	G	*amt, mdr, mtcq, nitro, rcq, rdq, rs, rweq, ts*	–	800.99

The frequencies of the different subsets obtained when the algorithm was run 100 times (for each value of K) are shown in the 5th column of Table 2. For example, for $K = 7$, the two subsets D and D' were obtained with frequencies 97% and 3%, respectively. For a given value of K, the most frequent subset will be preferred. Thus, for $K = 7$, the subset D must be kept, while D' is discarded, since the former has larger frequency than the latter. Furthermore, the five subsets A, B, C, D and E are the sufficient selected subsets for the five values of K, correspondingly. On the other hand, a subset with as few variables as possible must be preferred. Since subsets C and E have much more variables than subsets A, B and D, the former subsets can be discarded. Now, the remaining sets A, B and D have only three or two covariates, as well as the set F found in [16]. As mentioned earlier in this section, the spatial variation of diversity in mountain systems is an ongoing topic. Thus, these subsets *per se* can be of immense value to the experts interested in the diversity problem. The model in [16] can be used "to evaluate the performance" of the four subsets A, B, D and F by computing the BIC value for this GLM model in each of these subsets. The values so obtained are shown in the last column of Table 2. Among the sets A, B, D and F, the smallest BIC (793.82) is reached for set A. The set F proposed in [16] has a slightest larger BIC value. So, in terms of the selection criterion BIC, the KFOCI set A is "better" than the set F found in [16]. The subset G shown in the last row in Table 2 was obtained implementing the stepwise algorithm to the GLM model in [16] using BIC as the selection criterion. The BIC value of the "best" selected covariates is 800.99. Several comments are in order here, first the stepwise algorithm did not find the subset with smallest BIC, since the BIC value of subset A is smaller. Second, this set G has too many covariates. Third, although in [16], the set F was not obtained optimizing any selection criterion, the BIC value is remarkably close to the BIC value of the "best" model (with covariates in G) found by the stepwise algorithm. Also, set F has fewer covariates than the set G. So, the model in [16] is much more parsimonious than model G.

4 Conclusions

In this study, to exemplify the usefulness of KFOCI, a nonparametric procedure for variables selection, when an adequate parametric model is difficult to propose, was implemented to the dataset in [16], which has 30 covariates to study the variation in diversity in a mountain system. The KFOCI subsets A, B and D, shown in Table 2, have only a very few of the 30 covariates of the study. Based on a parametric model, in [16] the subset F (shown in Table 2) was selected. Like the KFOCI sets B and D, the subset F has only two covariates. The BIC values for the subsets A, B, D and F (shown in Table 2) are based on the parametric model in [16]. Also, the set G and its BIC value shown in this table are the solutions of the stepwise procedure implemented to the GLM model in [16] aimed to minimize the BIC criterion. Based on this model, the KFOCI set A is remarkably good. It has few covariates and has the smallest BIC value. Regarding the set F, although its BIC value is slightly larger than the BIC value of G, it has many more variables than F. Thus, the set F obtained in [16] is quite good. It is worth mentioning that the variables in both sets A and F are climate variables. Therefore, climate variables appear to be more important than the other factors in explaining diversity variation in mountain systems.

For the dataset in [16], the availability of a parametric model was extremely useful to select the "best" KFOCI subset. If this algorithm were implemented in a new study where a parametric model was not available, it will not be possible to compare the KFOCI subsets that may result, and clearly, this is a drawback of the KFOCI algorithm. In future work, another nonparametric variable selection procedure, such as the *model-X knockoffs* algorithm [6, 13, 20], could be implemented in this study.

References

1. Akaike, H.: A new look at the statistical model identification. IEEE transactions on automatic control **19**(6), 716–723 (1974)
2. Arellano, G., Tello, J.S., Jørgensen, P.M., Fuentes, A.F., Loza, M.I., Torrez, V., Macía, M.J.: Disentangling environmental and spatial processes of community assembly in tropical forests from local to regional scales. Oikos **125**(3), 326–335 (2016)
3. Azadkia, M., Chatterjee, S.: A simple measure of conditional dependence. The Annals of Statistics **49**(6), 3070–3102 (2021)
4. Benner, J., Vitousek, P., Ostertag, R.: Nutrient cycling and nutrient limitation in tropical montane cloud forests. Tropical montane cloud forests: science for conservation and management pp. 90–100 (2010)
5. Burnham, K.P., Anderson, D.R.: Model selection and multimodel inference: a practical information-theoretic approach. Springer (2002)
6. Candes, E., Fan, Y., Janson, L., Lv, J.: Panning for gold:'model-x'knockoffs for high dimensional controlled variable selection. Journal of the Royal Statistical Society Series B: Statistical Methodology **80**(3), 551–577 (2018)
7. Dalling, J.W., Heineman, K., González, G., Ostertag, R.: Geographic, environmental and biotic sources of variation in the nutrient relations of tropical montane forests. Journal of Tropical Ecology **32**(5), 368–383 (2016)

8. Gretton, A., Borgwardt, K.M., Rasch, M.J., Schölkopf, B., Smola, A.: A kernel two-sample test. The Journal of Machine Learning Research **13**(1), 723–773 (2012)
9. Guo, Q., Kelt, D., Sun, Z., Liu, H., Hu, L., Ren, H., Wen, J.: Global variation in elevational diversity patterns. sci rep 3: 3007 (2013)
10. Holland, J.H.: Complexity: A very short introduction. OUP Oxford (2014)
11. Huang, Z.: KPC: Kernel Partial Correlation Coefficient (2022). URL https://CRAN.R-project.org/package=KPC. R package version 0.1.2
12. Huang, Z., Deb, N., Sen, B.: Kernel partial correlation coefficient — a measure of conditional dependence. Journal of Machine Learning Research **23**(216), 1–58 (2022). URL http://jmlr.org/papers/v23/21-493.html
13. Kormaksson, M., Kelly, L.J., Zhu, X., Haemmerle, S., Pricop, L., Ohlssen, D.: Sequential knockoffs for continuous and categorical predictors: With application to a large psoriatic arthritis clinical trial pool. Statistics in Medicine **40**(14), 3313–3328 (2021)
14. Körner, C.: The use of 'altitude'in ecological research. Trends in ecology & evolution **22**(11), 569–574 (2007)
15. Kuhn, M., Johnson, K., et al.: Applied predictive modeling, vol. 26. Springer (2013)
16. Martínez-Camilo, R., González-Espinosa, M., Ramírez-Marcial, N., Cayuela, L., Pérez-Farrera, M.Á.: Tropical tree species diversity in a mountain system in southern mexico: local and regional patterns and determinant factors. Biotropica **50**(3), 499–509 (2018)
17. McCain, C.M.: Could temperature and water availability drive elevational species richness patterns? a global case study for bats. Global Ecology and biogeography **16**(1), 1–13 (2007)
18. Sanders, N.J., Rahbek, C.: The patterns and causes of elevational diversity gradients. Ecography **35**(1), 1 (2012)
19. Schwarz, G.: Estimating the dimension of a model. The annals of statistics pp. 461–464 (1978)
20. Vásquez, A.R., Márquez Urbina, J.U., González Farías, G., Escarela, G.: Controlling the false discovery rate by a latent gaussian copula knockoff procedure. Computational Statistics **39**(3), 1435–1458 (2024)
21. Wang, X., Fang, J., Sanders, N.J., White, P.S., Tang, Z.: Relative importance of climate vs local factors in shaping the regional patterns of forest plant richness across northeast china. Ecography **32**(1), 133–142 (2009)
22. Watkins Jr, J.E., Cardelús, C., Colwell, R.K., Moran, R.C.: Species richness and distribution of ferns along an elevational gradient in costa rica. American Journal of Botany **93**(1), 73–83 (2006)

Antisocial Behavior in Young Mexicans. Lifestyle Assessment and Risk Factors Detection

María Teresa Guerrero-San Vicente and Carlos Cuevas-Covarrubias

Abstract We present an original statistical model for assessing the risk of incurring in criminal conducts among young people in Mexico. It is based on supervised classification and reduction of dimensionality. The information is taken from a statistical survey with national coverage by the National Institute of Statistics in Mexico. Men and women are analyzed separately. The main contribution of this work is a classification criterion that expresses risk as a probability given some variables related to each person's lifestyle. The resulting model is accurate, informative, and interpretable.

Keywords CHAID · PCA · Logistic regression · Juvenile delinquency

1 Introduction

A major problem that Mexico faces at the moment is its increasing crime rates. During the last two decades, Mexican society has experienced an insecurity crisis affecting different aspects of lifestyle and life quality. During this period of time, an important phenomenon has been observed: the preoccupying participation of young people on criminal activities. In 2014, the federal government organized a statistical survey in order to assess the social health and lifestyle of young Mexican people in the must dangerous communities in the country. This survey has national coverage and measures different aspects of lifestyle, family dynamics, personal habits, and attitudes toward life. We present an interesting work of statistical analysis and modeling of different variables measured on this survey. The main contribution of this article is the presentation of a screening linear model to evaluate the risk of incurring in criminal conducts given different variables related to lifestyle and family dynamics. This model expresses risk as a probability of being arrested.

M. T. Guerrero-San Vicente (✉) · C. Cuevas-Covarrubias
Anáhuac University, Huixquilucan Estado de México, Mexico
e-mail: teresa.guerrero@anahuac.mx

J. A. Christen et al. (eds.), *Statistics, Society and Environment*, Springer Proceedings in Mathematics & Statistics 479, https://doi.org/10.1007/978-3-031-78401-9_14

We apply principal components analysis combined with logistic regression. The objective is to detect significant risk factors, and to build a linear model in order to assess risk. The final result is a risk score with high screening power. Men and women are analyzed separately to have a better understanding of the interaction of those factors related to criminal conducts.

2 Crime in Mexico

According to the *Global Study on Homicide* conducted by the United Nations Office on Drugs and Crime (UNDOC) [13] in 2021, the global homicide rate at that time was 5.8 per 100,000 inhabitants. In Mexico, the National Office for Public Security [12] (SESNSP[1]) reported a domestic homicides rate of 19.26 per 100,000 inhabitants in 2023. Figure 1 shows the trend of national crime rates from 2015 to 2023.

In Mexico, crime is strongly related to drug cartels activity. The US Council of Foreign Relations in its document *Criminal violence in Mexico* [1] explains that since the decade of 1980, Mexican drug dealers took regional areas to establish networks and trafficking routes. In 2006, President Felipe Calderón declared war

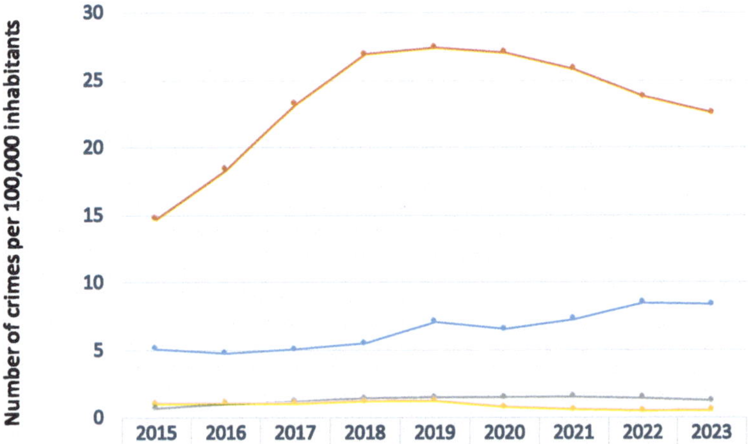

	2015	2016	2017	2018	2019	2020	2021	2022	2023
Homicide	14.74	18.37	23.27	26.92	27.43	27.05	25.87	23.81	22.61
Feminicide	0.69	1.03	1.21	1.44	1.5	1.5	1.55	1.47	1.27
Kidnaping	1.08	1.13	1.12	1.24	1.29	0.82	0.63	0.56	0.61
Extrosion	5.13	4.78	5.06	5.5	7.11	6.56	7.29	8.48	8.36

Fig. 1 Mexico: crime rates, 2015–2023

[1] Secretariado Ejecutivo del Sistema Nacional de Seguridad Pública.

to criminal organizations using military force. Since then, violence and crime have increased and now constitute a major problem in the country.

The participation of Mexican youth in organized crime is a complex problem that requires a multidisciplinary approach. Identifying and understanding different factors contributing to this phenomenon are essential in developing strategies to prevent criminal and antisocial behavior in young people. In 2014, Mexico's National Institute of Statistics and Geography (INEGI) published the results of the Social Cohesion Survey for the Prevention of Violence and Crime (ECOPRED) [4]. This survey has national coverage, involving 47 selected cities according to the National Program for the Social Prevention of Violence and Crime (PNPSVD). The sample size is $n = 40{,}366$ people aged between 12 and 29; 51% of them being men. All of them live in freedom at home with their families. ECOPRED was the first and last survey of its kind. Although it contains information 10 years old, it represents the most recent and complete source of information about violence and crime among young people in the country. The analysis presented in this work is made directly from the sample, and no statistical projections to the population are made.

3 Exploratory Data Analysis

Statistics reported in the literature indicate that crime is more prevalent among men; however, women with certain profiles have a higher risk of incurring in antisocial behavior and criminal conducts. These women tend to initiate their criminal activity at a remarkable young age. A deeper discussion of this fact can be found in Guerrero-San Vicente et al. [2], where the same survey is analyzed. We now focus on the answer to the question: *Have you ever had any conflict with the police (have you ever been arrested)?* given by youngsters responding ECOPRED. Throughout our discussion, those individuals who declared being arrested at least once in their lives will be referred to as *positives*. In a similar way, the group of people who have been arrested will be referred to as the *positive population*. The rest will be called *negatives* and *negative population*. Table 1 shows the prevalence of conflicts with the police for men and women in three different age groups. It is clear that men have higher prevalences than women; prevalence in the male population is in average five times higher. Clearly, to be a man is an important risk factor itself. It is also evident that the prevalence increases with age for the male group. In contrast, age doesn't seem to have an effect on the female group.

A Chi-square test confirms the effect of age on the male population, where the risk of incurring on crime or antisocial behavior increases through time. This is shown of Fig. 2.

Given the differences among men and women, we decided to analyze each population separately. The results of these analyses are shown in the next section.

Table 1 Arrest frequencies per age segment and gender

Age group	Men	Women
12–17	5.84%	1.94%
18–23	12.35%	2.01%
24–29	14.38%	2.03%
Total	9.77%	1.98%

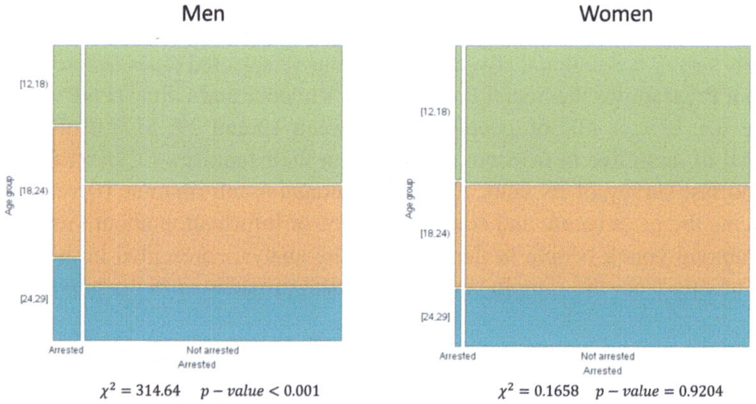

Fig. 2 ECOPRED 2014: Effect of age on criminal conducts

4 Statistical Analysis and Modeling

Guerrero-San Vicente et al. [2] present a **CHAID** regression tree for each gender. This allows us to identify important risk factors for both men and women (see Figs. 3 and 4). The segmentation given on each tree was included in logistic regression models for screening and risk assessment. The resulting models are accurate and interpretable for both genders, with areas under the ROC curves of 0.86 and 0.88 for a men and women respectively. Here, principal components analysis is used to explain the structure of the data and to reduce the number of predictors. The first principal component resulted to be a good discriminant function for both genders with an area under the ROC curve of 0.82 and 0.87 for men and women respectively. The first three principal components combined in a logistic regression model formed a better risk score with areas under the ROC curves of 0.83 and 0.88 for men and women respectively.

4.1 CHAID Analysis

According to Ritschard [11], CHAID[2] algorithm was introduced by Kass [7]. It is widely used as a supervised statistical classification method. Milanovic et al.

[2] Chi-square Automatic Interaction Detector.

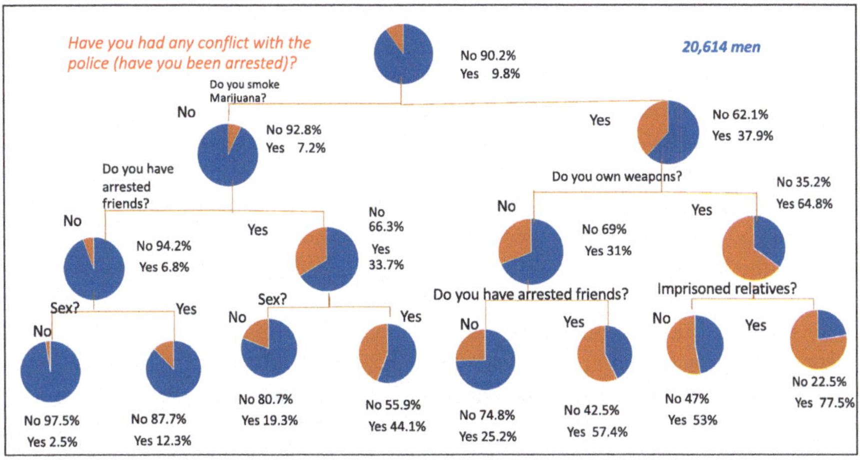

Fig. 3 Men CHAID analysis tree (taken from Guerrero-San Vicente et al. [2])

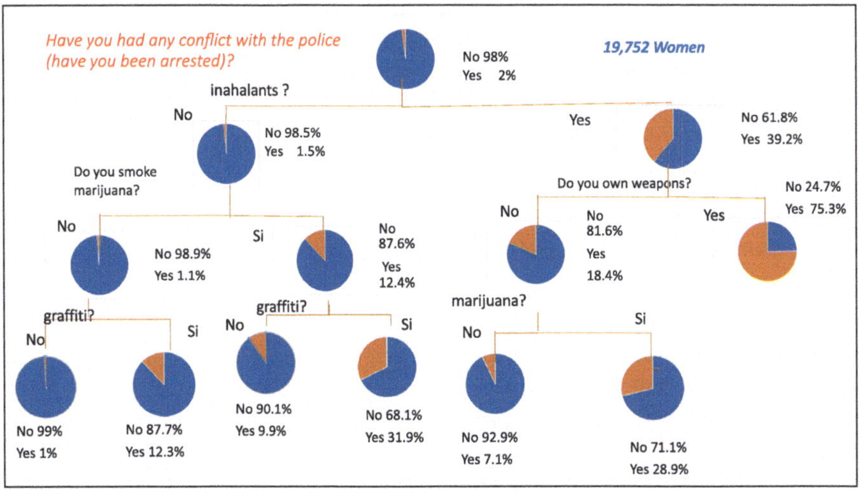

Fig. 4 Women CHAID analysis tree (taken from Guerrero-San Vicente et al. [2])

[10] comment that, because of its exploratory nature, CHAID analysis represents an ideal method for data structures discovering. It also helps to identify the most significant interactions between predictors and a target variable. The result is an informative segmentation of the sample expressed on a tree diagram. CHAID is a distribution-free algorithm. It was originally designed to analyze categorical and ordinal variables. However, it can also be applied to reveal interactions among continuous covariates. The CHAID analyses for men and women of ECOPRED are shown in Figs. 3 and 4. These two regression trees previously appeared in Guerrero-San Vicente et al. [2] where a new discriminant model for crime risk assessment is

presented. They reveal how marijuana consumption seems to be an important risk factor for the male population. Among those marijuana consumers, the risk of being arrested is 38% contrasting with a prevalence of 7% among non-consumers. The regression tree for the male population shows that those marijuana consumers who own weapons and have imprisoned relatives in their family constitute the group with the highest prevalence; 77.5% of men with this profile have been arrested at least once in their lives. On the other hand, CHAID analysis shows how men who are not marijuana consumers do not have arrested friends and do not have an active sexual life have the lowest prevalence in the male population where 97.5% have never been arrested. Similarly, the resulting tree for female population brings out that the use of inhalants as a drug, is one of the main risk factors for this population. The risk of being arrested increased up to 39.2% (almost 20 times) for those women who had used inhalants as a narcotic, contrasting with a prevalence of 1.5% for those who don't have this kind of practice. The prevalence is 75% for those women who had used inhalants as a drug and own weapons. Clearly, the use of inhalants as a drug in the female population represents an important risk factor. Guerrero-San Vicente et al. [2] use some of the variables in these CHAID trees and some of their interactions in order to build logistic regression models for screening and risk assessment. They present accurate logistic regression models where those variables are combined with some others previously mentioned in the literature. The results are accurate linear risk scores with large areas under their respective ROC curves: $AUC = 0.86$ for men and $AUC = 0.88$ for women.

4.2 Principal Components Analysis

Principal components analysis (PCA) is a feature extraction method based on the spectral decomposition of the covariance (correlation) matrix. The main objective is to represent a set of p random variables with a few noncorrelated factors called *Principal Components* (see Mardia et al. [8] pp. 212). This set of new variables summarize the data preserving as much information as possible. A specialized discussion about PCA for discrimination and classification can be found in Jolliffe [5] pp. 157–167. Here, we use PCA not only as a way to reduce dimensionality, but as a basis to build an interpretable discriminant function for crime risk assessment. ECOPRED reports more than 800 binary variables, indicating the presence or absence of different risk and protective factors related to the lifestyle of those who answered the survey. In order to reduce the number of variables in our analysis, cumulative risk indexes were built. Each index is the sum of several binary variables related to a specific risk factor. These indexes summarize the information about: domestic violence, lack of parental surveillance, family fights, imprisoned or arrested relatives and friends, risk conducts, and substance use. The resulting indexes are quantitative. In a second stage, a PCA was performed in order to reduce dimensionality even more. Considering that these cumulative indexes are different in nature, variance and orders of magnitude, their correlation matrix $\boldsymbol{\Sigma}$ was used

Table 2 First three principal components

	Men			Women		
Index	PC1	PC2	PC3	PC1	PC2	PC3
Domestic violence	0.43	0.22	0.56	0.40	0.29	−0.55
Substance use	0.54	−0.25	−0.22	0.54	−0.21	0.09
Risk conducts	0.51	−0.19	−0.45	0.54	−0.07	0.33
Imprisoned relatives	0.40	−0.03	−0.03	0.43	−0.05	0.34
Lack of surveillance	0.08	−0.67	0.62	0.16	−0.58	−0.65
Fights with relatives	0.32	0.64	0.21	0.21	0.73	−0.22
Area under ROC curve	0.82	0.59	0.59	0.87	0.67	0.60

as the basis for the PCA. Table 2 shows the coefficients of the first three Principal Components. In both cases (men and women), the first Principal Component has positive coefficients. The four most important variables in this first component are: domestic violence, substance use, risk conducts and imprisoned or arrested relatives or friends. The second Principal Component, shows a contrast between lack of parental surveillance and family fights. The meaning of the third principal component is slightly different between men and women. In the female case: it contrasts lack of parental surveillance and domestic violence with criminal practice and arrested/imprisoned relatives and friends. In the male case, it contrasts lack of parental surveillance and domestic violence with risk conducts only. It is worth noticing that the first Principal Component is a linear discriminant function itself with an area under its ROC curve of $AUC = 0.87$ for women and $AUC = 0.82$ for men (see Table 2). In this way, PCA can be effectively applied for screening of criminal behavior. The first three components were then summarized in a logistic regression model with an area under its ROC curve of 0.83 for men and 0.88 for women. Both regression models are presented in the next section.

Figure 5 shows how the first 2 Principal Components discriminate positives from negatives. Figures 6 and 7 illustrate how the first and fourth quadrants of the Cartesian plane have a much larger concentration of positives. Therefore, the interpretation of each quadrant in terms of the first two principal components may offer a risk profile for the individuals it contains. Further more, a logistic regression model may summarize the information in a single screening function expressing risk as a probability. As mentioned before, the first principal component is a general risk measure for both men and women (the larger its value, the higher the risk). The second component contrasts lack of parental surveillance vs family fights and domestic violence (see coefficients in Table 2). For both male and female, large values of the second component suggest frequent fights of the respondent with members of his family and domestic violence with a strong parental surveillance; on the other hand, negative values of the second component indicate lack of parental supervision but with rare fights with relatives and no domestic violence. The risk profile in quadrant one is: *young people living in a bad general environment with domestic violence and frequent fights at home, but with a protective parental*

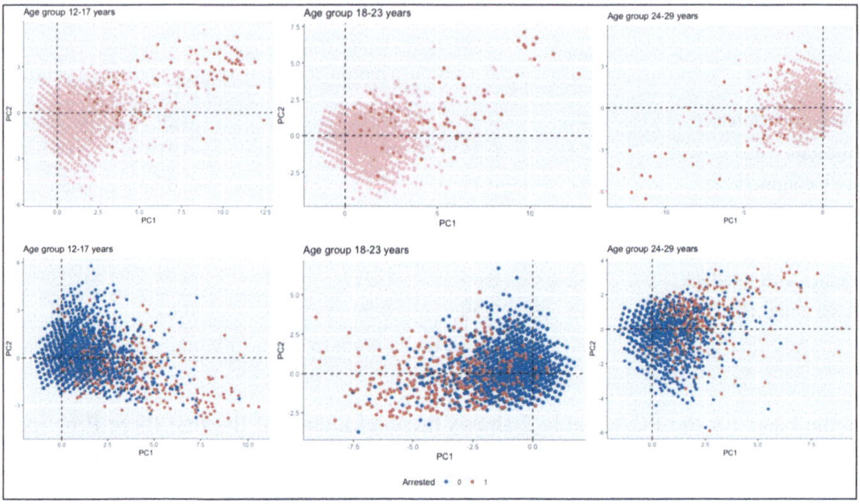

Fig. 5 PCA scatter plots: women in top row, men in bottom row

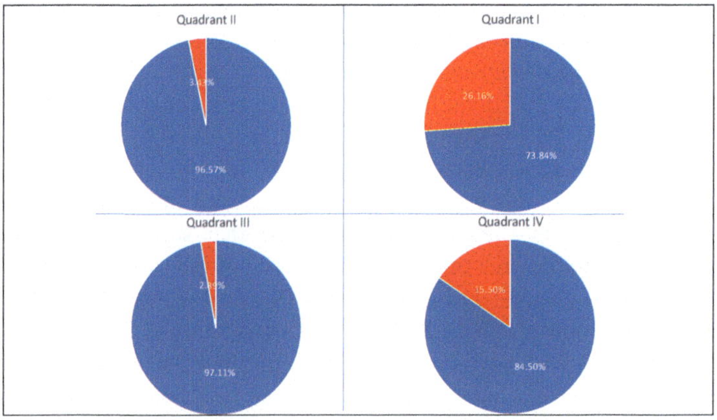

Fig. 6 Men Cartesian quadrants distribution

surveillance. The risk profile in quadrant four is: *young people living in a bad general environment with poor parental surveillance but with rare fights at home and low levels of domestic violence*.

In order to explore the effects of age on the risk of incurring in criminal conducts, the total sample was divided in three different age groups; then, a PCA was applied for each group separately. Although the result is practically the same as the one obtained with the complete sample, the smaller sample sizes facilitate the visualization of the PCA scores on the Cartesian plane. It is clear that those

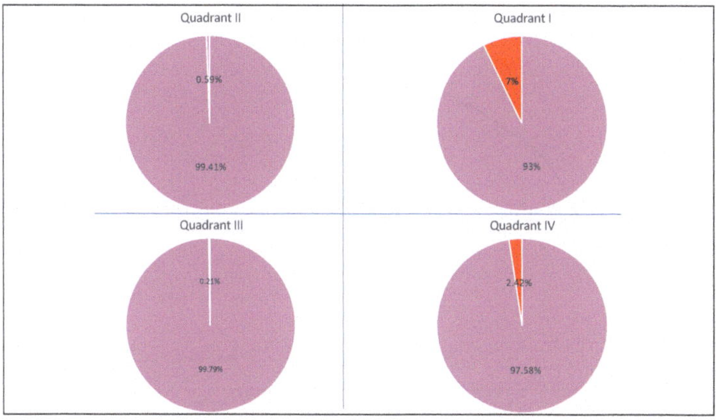

Fig. 7 Women Cartesian quadrants distribution

individuals who have been arrested are separated from those who have never being arrested. This confirms that Principal Components in this example constitute two linear risk scores useful for our screening purpose.

4.3 *Logistic Regression*

Logistic regression is a supervised classification method. Given a set of covariates, it expresses posterior probabilities in terms of a linear function. It is frequently applied to discriminate individuals from two possible categories: positives and negatives. Consider a random vector $X : \Omega \rightarrow \mathbb{R}^p$ defined on a sample space Ω, which is partitioned into two subsets: Ω_0 and Ω_1. Given $x = X(\omega)$, the objective is to infer whether $\omega \in \Omega_0$ or $\omega \in \Omega_1$. The model expresses posterior probabilities according to:

$$\theta(\boldsymbol{X}) = P(Y = 1|\boldsymbol{X} = \boldsymbol{x}_i) = \frac{\exp(\beta_0 + \boldsymbol{\beta}^T \boldsymbol{x}_i)}{1 + \exp(\beta_0 + \beta^T \boldsymbol{x}_i)}$$

Interesting discussions about logistic regression for discrimination can be found in Hastie et al. [3] pp. 119–127, Izenman [6] pp. 250–253, and McCullagh et al. [9] pp. 107–138.

For both male and female samples, different logistic regression models were fitted. In a previous work by Guerrero-San Vicente et al. [2], categorical binary variables and interactions detected with the CHAID were applied. In contrast, here we use as predictors the first three Principal Components analyzed in the previous subsection. Both logistic regression models are accurate, with areas under the ROC curve of $AUC = 0.83$ for men and $AUC = 0.88$ for women (see Fig. 8). If the minimax threshold is applied, this model shows a sensitivity of 0.80 for women and 0.76 for men, and specifities of 0.80 for women and 0.75 for men (see Fig. 8).

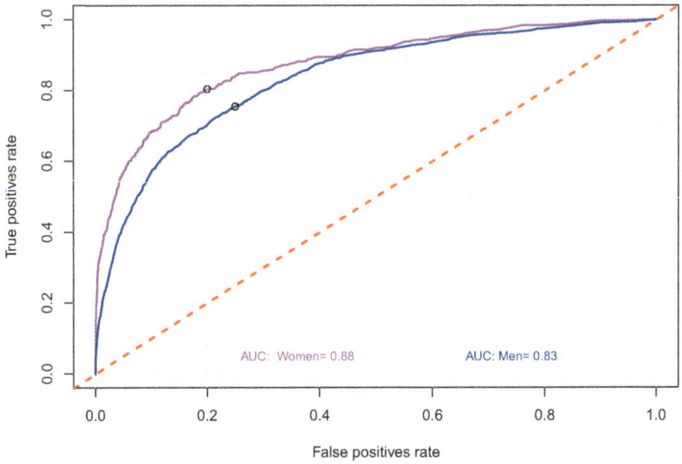

Fig. 8 Logistic regression models: areas under ROC curve

Table 3 Logistic regression models: comparison based on AUC

Model	Men	Women
Based on PCA	0.83	0.88
Based on CHAID	0.86	0.88

Table 4 Logistic regression coefficients the first three PCs

Predictor	Men	Women
$\hat{\beta}_0$	−2.65	−4.63
PC1	0.76	0.69
PC2	−0.23	−0.22
PC3	0.18	0.10

Contrasting with the screening model previously presented in Guerrero-San Vicente et al. [2], our logistic regression model is built upon quantitative variables. This model based on PCA is easily interpretable and offers complementary information. Both approaches (CHAID and PCA) are accurate with large areas under their ROC curves (see Tables 3 and 4).

Although both approaches are accurate, the PCA approach is more interpretable and more easily implemented on any computer.

5 Conclusion

ECOPRED is an important source of information about crime and its causes in Mexico. Nevertheless, the lack of references to this survey in the literature suggests that ECOPRED has not been exploited in its full potential. We consider that public policy in Mexico should take empirical evidence like ECOPRED into account. The main contribution of this chapter is a multivariate analysis that reveals the importance of certain risk factors related to the way of life of young Mexican people and a accurate logistic regression model explaining the relation between risk conducts, lifestyle, and family dynamics with potential crime in young Mexicans.

Contrasting with other works in the literature, we analyzed the information for men and women separately. It is clear that: potential crime and conducts of risk are more prevalent in the male population; however, the final screening model resulted to be more accurate for the female population.

The PCA confirms that substance use is a risk factor of major importance for both genders. It also describes the role of family dynamics as a way to prevent crime among young people; the lowest proportion of people arrested was observed among respondents with a low level of risk conducts and a home without violence. This conclusion is an indication to design public policies promoting family values together with healthy lifestyles.

To live in a family with imprisoned members is also a risk factor. A good recommendation would be to facilitate psychological support to families in this situation. An educational campaign informing parents about the potential influence of comrades and friends is also recommendable.

The fact that the logistic regression model is accurate indicates that crime is something preventable among young people and induces an optimistic view to the future.

Acknowledgments Both authors express their gratitude to the Anáhuac University and FICSAC from Universidad Iberoamericana who kindly sponsored this project.

References

1. Council of Foreign Relations (2023); Criminal Violence in Mexico, Global Conflict Tracker. Available at: https://www.cfr.org/global-conflict-tracker/conflict/criminal-violence-mexico
2. Guerrero-San Vicente, M.T. Cuevas-Covarrubias, C. (2022); *Conductas delictivas en jóvenes mexicanos, detección de factores de riesgo y modelación estadística* in Martínez-Lanz, P. (ed.) *Delincuencia en México, Revisión interdisciplinar*, Porrúa and Universidad Anáhuac.
3. Hastie, T. Tibshirani, R. Friedman, J. (2009); *The elements of statistical learning. Data mining, inference and prediction*, Springer. DOI: https://doi.org/10.1007/978-0-387-84858-7
4. Instituto Nacional de Estadística y Geografía, (2015). Encuesta de Cohesión Social para la Prevención de la Violencia y la Delincuencia (ECOPRED) 2014 *INEGI* Available at: https://www.inegi.org.mx/programas/ecopred/2014/
5. Jolliffe, I.T. (2002); *Principal Component Analysis*, Springer. ISBN: 978-0-387-22440-4. DOI: https://doi.org/10.1007/b98835
6. Izenman A.J. (2016); *Modern multivariate statistical techniques: regression, classification and manifold learning*, Springer. ISBN 978-1-4939-3832-2, DOI: https://doi.org/10.1007/978-0-387-78189-1
7. Kass, G. D. (1980); An exploratory technique for investigating large quantities of categorical data, *Journal of the Royal Statistical Society Series C (Applied Statistics)*, Vol. 29 No. 2, pp 119–127, DOI: https://doi.org/10.2307/2986296

8. Mardia K.V., Kent J.T., Bibby J.M. (1979); Multivariate Analysis, Academic Press. DOI: https://doi.org/10.1002/bimj.4710240520
9. McCullagh P., Nelder J.A. (1989) *Generalized Linear Models* Second edition, Chapman & Hall/CRC. DOI: http://dx.doi.org/10.1007/978-1-4899-3242-6
10. Milanovic, M.,Stamenkovic, M. (2016). CHAID Decision Tree: Methodological Frame and Application, Economic Themes, 54(4), 563–586. Available at: https://doi.org/10.1515/ethemes-2016-0029
11. Ritschard, G. (2014); *CHAID and Earlier Supervised Tree Methods* in McArdle, J. J., & Ritschard, G.; *Contemporary Issues in Exploratory Data Mining in the Behavioral Sciences*, Routledge. DOI: https://doi.org/10.4324/9780203403020
12. Secretariado Ejecutivo del Sistema Nacional de Seguridad Pública (SESNSP), (2024); *Reporte Estatal de Incidencia delictiva del Fuero Común, nueva metodología.* Available at: https://www.gob.mx/sesnsp/acciones-y-programas/incidencia-delictiva-del-fuero-comun-nueva-metodologia?state=published
13. United Nations Office On Drogs and Crime (UNDOC)(2023); Global study on homicide 2023, UNDOC Research, Vienna, United Nations, New York, Available at: https://www.cdeunodc.inegi.org.mx/index.php/inicio-publicaciones/

Author Index

J. A. Christen et al. (eds.), *Statistics, Society and Environment*, Springer Proceedings in Mathematics & Statistics 479, https://doi.org/10.1007/978-3-031-78401-9

Subject Index

J. A. Christen et al. (eds.), *Statistics, Society and Environment*, Springer Proceedings in Mathematics & Statistics 479, https://doi.org/10.1007/978-3-031-78401-9

The manufacturer's authorised representative in the EU is Springer
Nature Customer Service Centre GmbH, Europaplatz 3, 69115 Heidelberg,
Germany. If you have any concerns regarding our products, please
contact ProductSafety@springernature.com

Printed and bound by CPI Group (UK) Ltd, Croydon, CR0 4YY
01/10/2026
02225960-0004